AF526226

Springer Series in **Materials Science** 20

Edited by J. Peter Toennies

Springer Series in **Materials Science**

1 **Chemical Processing with Lasers**
By D. Bäuerle

2 **Laser-Beam Interactions with Materials**
Physical Principles and Applications
By M. von Allmen

3 **Laser Processing of Thin Films and Microstructures**
Oxidation, Deposition and Etching of Insulators
By I. W. Boyd

4 **Microclusters**
Editors: S. Sugano, Y. Nishina, and S. Ohnishi

5 **Graphite Fibers and Filaments**
By M. S. Dresselhaus, G. Dresselhaus, K. Sugihara, I. L. Spain, and H. A. Goldberg

6 **Elemental and Molecular Clusters**
Editors: G. Benedek, T. P. Martin, and G. Pacchioni

7 **Molecular Beam Epitaxy**
Fundamentals and Current Status
By M. A. Herman and H. Sitter

8 **Physical Chemistry of, in and on Silicon**
By G. F. Cerofolini and L. Meda

9 **Tritium and Helium-3 in Metals**
By R. Lässer

10 **Computer Simulation of Ion–Solid Interactions**
By W. Eckstein

11 **Mechanisms of High Temperature Superconductivity**
Editors: H. Kamimura and A. Oshiyama

12 **Dislocation Dynamics and Plasticity**
By T. Suzuki, S. Takeuchi, and H. Yoshinaga

13 **Semiconductor Silicon**
Materials Science and Technology
Editors: G. Harbeke and M. J. Schulz

14 **Graphite Intercalation Compounds I**
Structure and Dynamics
Editors: H. Zabel and S. A. Solin

15 **Crystal Chemistry of High T_c Superconducting Copper Oxides**
By B. Raveau, C. Michel, M. Hervieu, and D. Groult

16 **Hydrogen in Semiconductors**
By S. J. Pearton, M. Stavola, and J. W. Corbett

17 **Ordering at Surfaces and Interfaces**
Editors: A. Yoshimori, T. Shinjo, and H. Watanabe

18 **Graphite Intercalation Compounds II**
Editors: S. A. Solin and H. Zabel

19 **Laser-Assisted Microtechnology**
By S. M. Metev and V. P. Veiko

20 **Microcluster Physics**
By S. Sugano

Satoru Sugano

Microcluster Physics

With 125 Figures

Springer-Verlag
Berlin Heidelberg New York
London Paris Tokyo
Hong Kong Barcelona
Budapest

Professor Dr. Satoru Sugano

Faculty of Science, Himeji Institute of Technology,
Kamigori-chyo, Ako-gunn 678-12, Japan and
Institute for Solid State Physics, University of Tokyo,
Roppongi, Minato-ku, Tokyo 106, Japan

Guest Editor:
Professor Dr. J. Peter Toennies

Max-Planck-Institut für Strömungsforschung,
Bunsenstrasse 10, W-3400 Göttingen, Fed. Rep. of Germany

ISBN 3-540-53926-3 Springer-Verlag Berlin Heidelberg New York
ISBN 0-387-53926-3 Springer-Verlag New York Berlin Heidelberg

This text was prepared using the PS™ Technical Word Processor

54/3140-543210 – Printed on acid-free paper

Preface

This book aims at providing graduate students and researchers with fundamental knowledge indispensable for entering the new field of "microclusters". Microclusters consisting of 10 to 10^3 atoms exhibit neither the properties of the corresponding bulk nor those of the corresponding molecule of a few atoms. The microclusters may be considered to form a new phase of materials lying between macroscopic solids and microscopic particles such as atoms and molecules, showing both macroscopic and microscopic features. However, research into such a new phase has been left untouched until recent years by the development of the quantum theory of matter.

The microscopic features of microclusters were first revealed by observing anomalies of the mass spectrum of a Na cluster beam at specific sizes, called magic numbers. Then it was experimentally confirmed that the magic numbers come from the shell structure of valence electrons. Being stimulated by these epoch-making findings in metal microclusters and aided by progress of the experimental techniques producing relatively dense, non-interacting microclusters of various sizes in the form of microcluster beams, the research field of microclusters has developed rapidly in these 5 to 7 years. The progress is also due to the improvement of computers and computational techniques, which have made it possible to perform ab initio calculations of the atomic and electronic structure of smaller microclusters, as well as to carry out computer simulations of their dynamics.

The field of microclusters is attracting the attention of many physicists and chemists (and even biologists!) working in both pure and applied research, as it is interesting not only from the fundamental point of view but also from the viewpoint of applications in electronics, catalysis, ion engineering, carbon-chemical engineering, photography and so on. At this stage of development, it is felt that an introductory book is required for beginners in this field, clarifying fundamental physical concepts important for the study of microclusters. This book is designed to satisfy such a requirement. It is based on series of lectures given to graduate students (mainly in physics) of the University of Tokyo, Kyoto University, Tokyo Metropolitan University, Tokyo Institute of Technology and Kyushu University in the period 1987-1990.

The book contains chapters on the definition of microclusters (Chap.1), dynamical and thermodynamical properties (Chap.2), the shell model and fission of metal clusters (Chap.3), ab initio calculations of alkali-, noble- and transition-metal clusters, and divalent and trivalent metal clusters (Chap.4), semiconductor clusters (Chap.5), rare-gas clusters (Chap.6), molecular clusters (Chap.7), and miscellaneous topics of synthetic chemistry, photographic latent-images, small metal clusters, mercury clusters and prospects of microcluster research (Chap.8). As mentioned already, the book is not a review article, so the author suggests that readers use the proceedings of related conferences[1] to find reference papers on individual matters, although representative papers concerning the fundamentals of microcluster physics are listed at the end of the book.

Recently, major progress in two areas of the field of microclusters has been noticeable. The first instance is related to the observation of magic numbers for large microclusters of 10^3 to 10^4 sodium atoms and the theoretical prediction of the existence of a super-shell for large metal clusters (Sect.8.5). The second is success in producing a fairly large number of C_{60} and C_{70} microclusters (Sect.5.1.1) and growing crystals consisting of these microclusters. This success seems to promise a future development of a new field of carbon chemistry.

Both experimental and theoretical studies of the fission of multiply charged metal microclusters are developing rapidly. Fission is induced by releasing the Coulomb repulsion energy of the charges confined in a small volume. In this sense, cluster fission is similar to nuclear fission and different from chemical reaction. It is felt that the energy transformation in the fission of multiply charged clusters may have potential applications, although the amount of energy involved is only of the same order as that in chemical reactions. Theoretical studies of symmetric fission performed recently in the author's research group by using the theory of shell corrections developed in nuclear physics are described in Chap.3.

The author is indebted to many people in his research group for help while preparing the book; in particular to Dr. Y. Ishii and Dr. A. Tamura for providing unpublished figures, and to Professor T. Yamaguchi for critical reading of the manuscript. The author expresses his sincere thanks to Professor J. Friedel for critical discussions on many topics included in this book during the author's stay in Paris and to Professor W. Kohn for stimulating discussions on the fission of multiply charged metal clusters while at the Institute of Theoretical Physics, University of California, Santa Barbara. The author also thanks Professor W.D. Knight, Professor R.S. Berry, Professor C. Bréchignac and Professor M.L. Cohen for illuminating discussions. The manuscript could not have been completed without the secretarial aid of Mrs. K. Fujii.

Himeji
February 1991 *S. Sugano*

[1] For example, E. Recknagel, O. Echt (eds.): Proc. 5th Int'l Symp. on Small Particles and Inorganic Clusters (Springer, Berlin, Heidelberg 1991). This book originally appeared as Volumes 19 and 20 of Z. Physik D.

Contents

1. What are Microclusters?

Classification of the fragments obtained by successive division of material is given according to their sizes. The particles called *fine particles* consisting of 10^3-10^5 atoms exhibit properties different from those of the material before division at low temperatures. They have to receive a statistical treatment, as we are unable to pick up the fine particles of a given shape: they have uncontrollable surface irregularity. On the other hand, the particles called *microclusters* consisting of 10-10^3 atoms show quantum-mechanical properties depending upon their shape like atoms and molecules. The microclusters of given shape and size can, in principle, be extracted and their properties can be measured. It is also discussed that microclusters exhibit, at finite temperatures, physical properties often encountered in macroscopic systems like liquids. This is due to the presence of a large number of low-lying metastable states.

1.1 Constituent Small Particles of Material

It looks quite natural that Greek and Indian philosophers have contemplated about successive division of materials and concluded the existence of constituent small particles to be no longer divisible, although the atomic theory of materials originated by J. Dalton is beyond our common imagination.

Let us again contemplate about successive division of materials like Greek and Indian philosophers but with some knowledge of the atomic theory, and raise naive questions. The first question to be brought up would be as follows. "How small could be the constituent particles still having properties as those of the material before division?" The second question immediately following the first one would be; "if these constituent particles of the smallest size were further divided, would the fragments be molecule-like or atom-like?"

We have also learned statistical mechanics besides quantum mechanics, and know that statistical mechanics should connect the macroscopic world of materials with the microscopic world of atoms and molecules. This knowledge makes us raise the third question as follows. "Is the world, the constituent particles of the smallest size belong to, just the beginning of the macroscopic one?" In other words, would the microscopic world suddenly emerge when the smallest constituent particles were fragmented? If the transition between the macroscopic and microscopic worlds was not so sudden, what kind of statistical mechanics would work in this range of the transition?

1.2 Division of Materials

Without responding to the questions raised in the previous section, let us first classify the fragments according to their sizes when successive division of a material is performed. It will soon become clear why the classification (Fig. 1.1) has to be adopted. In Fig. 1.1, three kinds of quantities are indicated to show the size; the number of constituent atoms, the radius, and the ratio of the number of inside atoms to that of surface atoms.

1.2.1 Fine Particles

Here we confine ourselves to metals, for simplicity. When we arrive at a fragment with a radius of the order of 100 Å by successive division of a metal, we see that the fragment exhibits properties different from those of the material before division at low temperatures. This was first pointed out by *Kubo* [1.1]. We shall call the fragments in this range of size "fine particles".

The main reason for the appearance of such a new phase is explained as due to the increase of the statistically averaged separation δ of the energy levels of a valence electron up to the order of 1 K (10^{-4} eV) when the fragment size is decreased down to this range of size. Here, the fragment contains 10^5 atoms (valence electrons) as depicted in Fig. 1.1, while the Fermi level, which is almost independent of the particle size (Sect. 3.3.2), would be of the order of 10 eV. Then, if one could assume a *homogeneous* statistical distribution of the energy levels, δ would be given as $(10\,\mathrm{eV}/10^5) = 10^{-4}$ eV ~ 1 K.

Now, let us assume the probability of finding an energy level in the region between x and x+dx to be P(x)dx, x being the unfolded level spac-

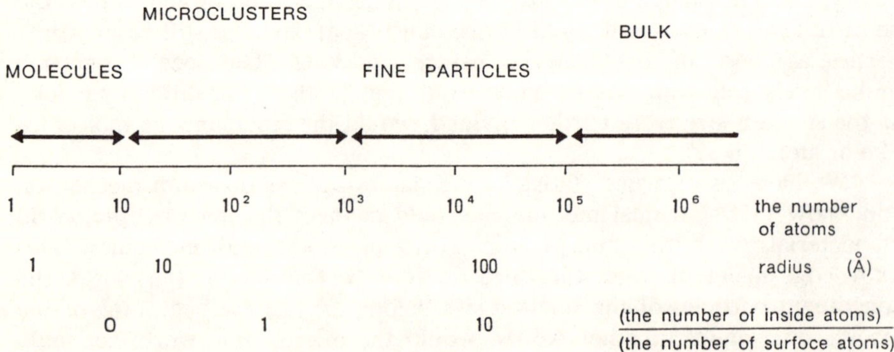

Fig. 1.1. Classification of the fragments according to their sizes obtained by successive division of material

ing defined as $x = \Delta\epsilon/\delta$, where $\Delta\epsilon$ is the energy spacing between the neighboring levels, being independent of energy position ϵ of the level because of the assumed homogeneity of the level distribution. In the absence of the spin-orbit interaction and a magnetic field, the application [1.2] of Wigner's random matrix theory [1.3] reveals that P(x) is given by Wigner distribution;

$$P(x) = \frac{\pi}{2} x e^{-\pi x^2/4} . \tag{1.1}$$

Note that this distribution gives $\langle x \rangle = 1$, $\langle x \rangle$ being the averaged of x. If one uses P(x) in (1.1) and calculates, for example, the electronic specific heat, one may show that it is proportional to T^2 for $kT \ll \delta$ [1.4] in contrast to T for the bulk.

The most important characteristic of the physics of fine particles comes from the statistical treatment of the distribution of electronic energy-levels, as found by the use of the random matrix theory. The statistical treatment is considered to be justified by the existence of uncontrollable surface irregularity: we are unable to pick up the particles of a given shape. This point is quite different from the physics of microclusters where surface boundary conditions play an essential role. In what follows, we shall give a simple example of computer-grown, two-dimensional metal particles with irregular surfaces, which shows the electronic energy levels of Wigner's distribution [1.5].

We ask a computer to grow particles of size N (the number of atoms contained) on the two-dimensional triangular lattice by using the following algorithm. To grow the particle of size N′+1 from that of size N′, an atom is added at one of the unoccupied sites neighboring the occupied sites with probability $Q(\xi)$ defined by

$$Q(\xi) = e^{\alpha\xi} \Big/ \sum_{\xi=1}^{z} e^{\alpha\xi} , \tag{1.2}$$

where ξ, z and α represent, respectively, the number of occupied sites around the site under consideration, the number of the nearest-neighbor lattice sites, and the parameter governing surface irregularity. The particle grown in such a way tends to have a smooth surface when the parameter α is large enough, while it has a rough surface when α is small. We define the surface irregularity R by

$$R = \frac{L^2}{48N} - 1 , \tag{1.3}$$

where L is the perimeter of the particle. In (1.3) integer 48 comes from such a normalization that R = 0 if the particle is a regular hexagon. We denote by (N, R) the ensemble of the particles containing N atoms with irregularity R. Some of the computer-grown particles are depicted in Fig.1.2.

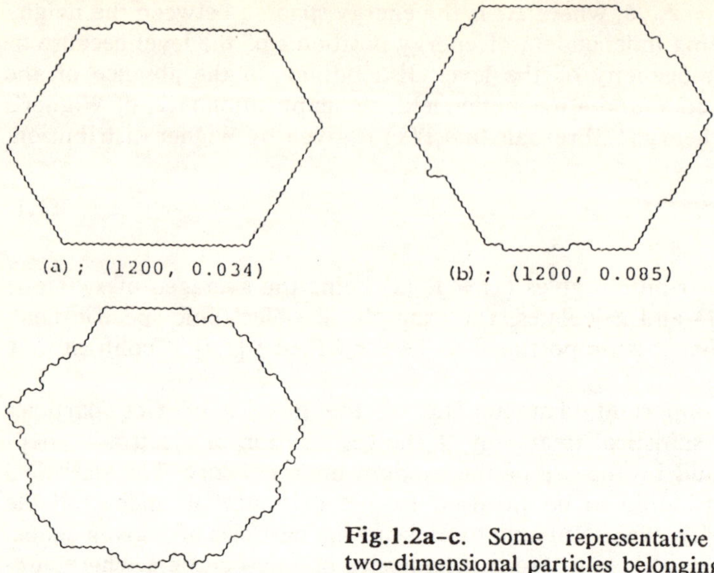

Fig.1.2a–c. Some representative computer-grown two-dimensional particles belonging to the ensembles (N, R) [1.5]

Once a particle is grown, we diagonalize the tight-binding Hamiltonian $\mathscr{H}$ of an electron in a particle as follows:

$$\mathscr{H} = - \sum_{\langle i,j \rangle} c_i^+ c_j \ , \tag{1.4}$$

where i and j in $\langle i, j \rangle$ are the nearest-neighbor sites of each other, and c_i^+ and c_i the creation and annihilation operators, respectively. We choose an energy scale in which the transfar integral is unity. Spins are ignored in our treatment. Examining the energy eigenvalues thus obtained for 600 particles in a fixed ensemble (N, R), we see that the ensemble average of the unfolded level spacing x is nearly independent of energy over the wide range around the middle of the energy band. The calculated distributions of the level spacing x around the middle of the energy band (Fermi level) are shown for a few ensembles in Fig.1.3. It is instructive to fit the calculated distributions with Brody distributions defined as

$$P_B(x;\omega) = (\omega + 1)\kappa(\omega)x^{\omega}\exp[-\kappa(\omega)x^{\omega+1}] \ , \tag{1.5}$$

with

$$\kappa(\omega) = \left[\Gamma\left(\frac{\omega + 2}{\omega + 1}\right)\right]^{\omega+1} ,$$

where $\Gamma(x)$ is the Γ-function of x. The Brody distribution is devised as an interpolation formula connecting the Wigner's distribution (ω=1) and Pois-

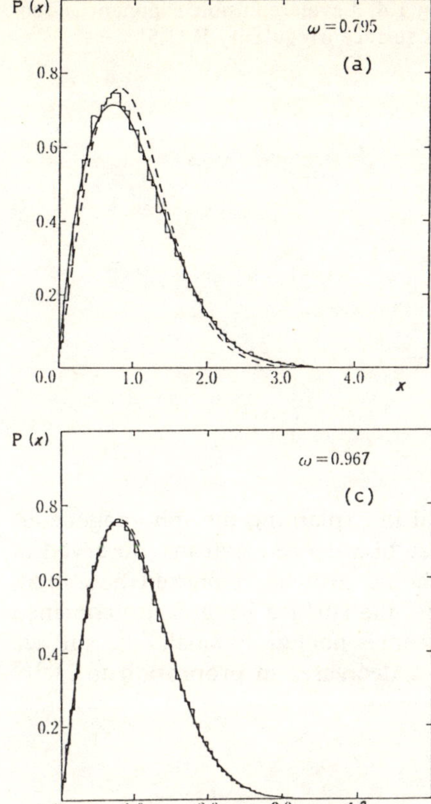

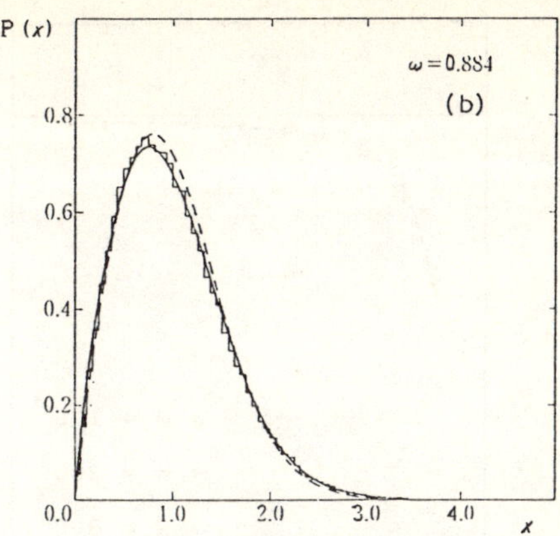

Fig.1.3. Distribution of the unfolded level spacing x for ensembles (300, 0.068), (300, 0.102), and (300, 0.210) in (a), (b), and (c), respectively. Solid curves are the Brody distributions $P_B(x,\omega)$ whose ω is indicated. Wigner's distributions are indicated by dashed curves for comparison [1.5]

son's distribution (ω=0). Since Poisson's distribution of the energy levels is expected if non-diagonal elements of the energy matrix inducing repulsion between the levels are neglected, the parameter ω may be regarded as an exponent to measure the level repulsion. In Fig.1.4, we plot ω determined by the fitting, as depicted in Fig.1.3, as a function of the irregularity parameter R. From the figure we may conclude that Wigner's distribution is applicable to the statistical treatment of the energy levels, in other words, the random matrix theory is valid, as long as fine particles have enough surface irregularity, R $\geq$ 0.2; a similar conclusion may be obtained in the three-dimensional case, too [1.5].

In this subsection, we have discussed, at length, the validity of the statistical treatment of electronic energy levels of fine particles with uncontrollable surface irregularity. We have not, however, discussed explicitly its validity when the particle size becomes very small. In our treatment, the level statistics breaks down for very small particles due to the shortage of the members of the ensemble. In the next subsection, we shall demonstrate that, in the size range of mciroclusters (Fig.1.1), the statistical treatment is invalid and the quantum mechanical treatment of the boundary condition

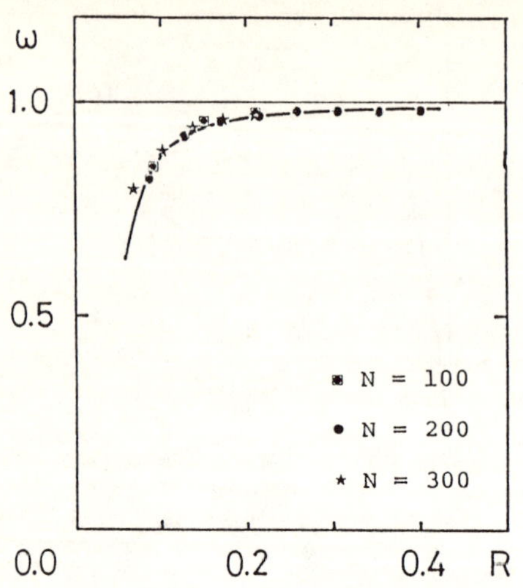

Fig.1.4. Level-repulsion exponent ω versus surface irregularity R [1.5]

coming from the cluster shape is essential in explaining the inhomogeneous level distribution, called the shell structure of valence electrons, observed in the mass spectra of metal clusters. It should also be remarked here that, when the particle size becomes very large, the surface irregularity confined to a few atomic layers at the surface becomes negligibly small: the surface irregularity of three-dimensional particles decreases in proprotion to $N^{-1/3}$ if N is very large [1.5].

1.2.2 Microclusters

When we arrive at the fragment called *microcluster* with the radius of the order of 10 Å by dividing further fine particles, we see that we have to use physics different from that for fine particles. The essential difference comes from the theoretical postulate, partly supported by experiments, that the microclusters of given shape and size can, in principle, be extracted and their properties can be measured although this kind of measurements is impossible for fine particles. This postulate may be justified by considering the fact that the clusters of a given regular shape are very stable as compared with those of the other shapes, the number of which is rather small. In contrast to this fact, the fine particles of different shapes and a fixed size forming a big ensemble to allow statistical treatments are nearly degenerate in energy. This makes impossible the extraction of the fine particles of a given shape.

Clear-cut evidence has been obtained for microclusters of alkali [1.6] and noble [1.7] metal elements in the form of a cluster beam to have a nearly spherical shape at the size of the so-called *magic numbers*. A magic number means a specific size N where anomalies of abundance in the mass spectra are found, indicating the microclusters of these sizes to be relatively stable as compared with those of the neighboring sizes. As an example, we show the mass spectrum of Na cluster beam in Fig.1.5. The beam is produced by the adiabatic expansion of the heated Na and Ar mixed gas

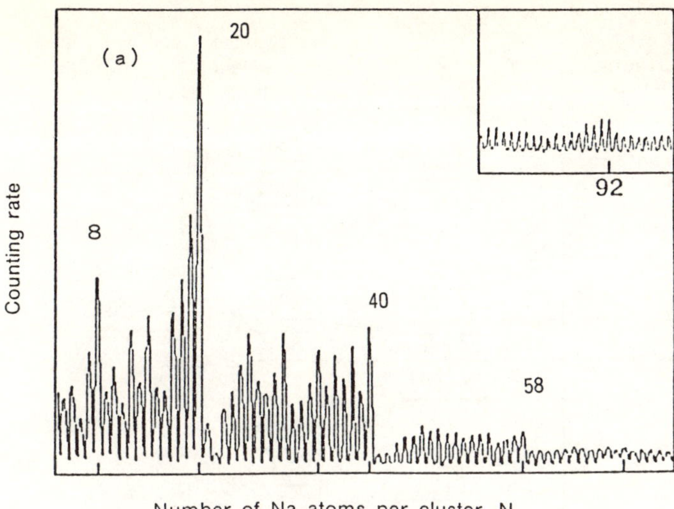

Fig. 1.5. Mass spectrum of Na cluster beam with magic numbers indicated [1.6]

through a nozzle. The Na clusters in the beam are photoionized, mass analyzed by a quadrupole mass analyzer, and finally detected by an ion-detection system. Detailed examinations of the experiment verify that the mass spectrum thus observed reflects that of neutral clusters originally produced by the jet expansion. The anomalies of abundance at the sizes N being 8, 20, 40, 58 and 92 (Fig. 1.5) are regarded as the magic numbers of neutral Na clusters. In what follows, we shall show that these magic numbers are associated with the shell structure of valence electrons moving independently in a spherically symmetric effective potential.

Figure 1.6 is taken from the classic book on the nuclear shell structure [1.9], showing the energy-level diagram of a particle, and electron in our case, in three-dimensional spherically symmetric potentials of the following types; the square well with an infinite wall (right column), the harmonic oscillater (left column), and the intermediate (central column) obtained by rounding the square well or flattening the bottom of the harmonic-oscillater potential. In the square well and the intermediate potentials, the energy levels are classified into the $n\ell$ ($n = 1,2,\ldots$, $\ell = s,p,d,\ldots$) shells with degeneracy $(2\ell+1)$, although an additional degeneracy exists in the harmonic oscillator potenial. Here, ℓ is the angular momentum and n, being different from the principal quantum number of a hydrogen atom, are integers only to distinguish different shells with the same ℓ. In the figure, the number in square brackets shown above each level indicates the number of total electrons when the relevant level is filled up. We immediately notice that some of these shell-closing numbers correspond to the magic numbers in Fig. 1.5. From this fact, we conclude that, dividing Na cluster of size N into N valence electrons and N ion cores, the valence electrons have the shell structure coming from the individual motion in a spherically-symmetric effec-

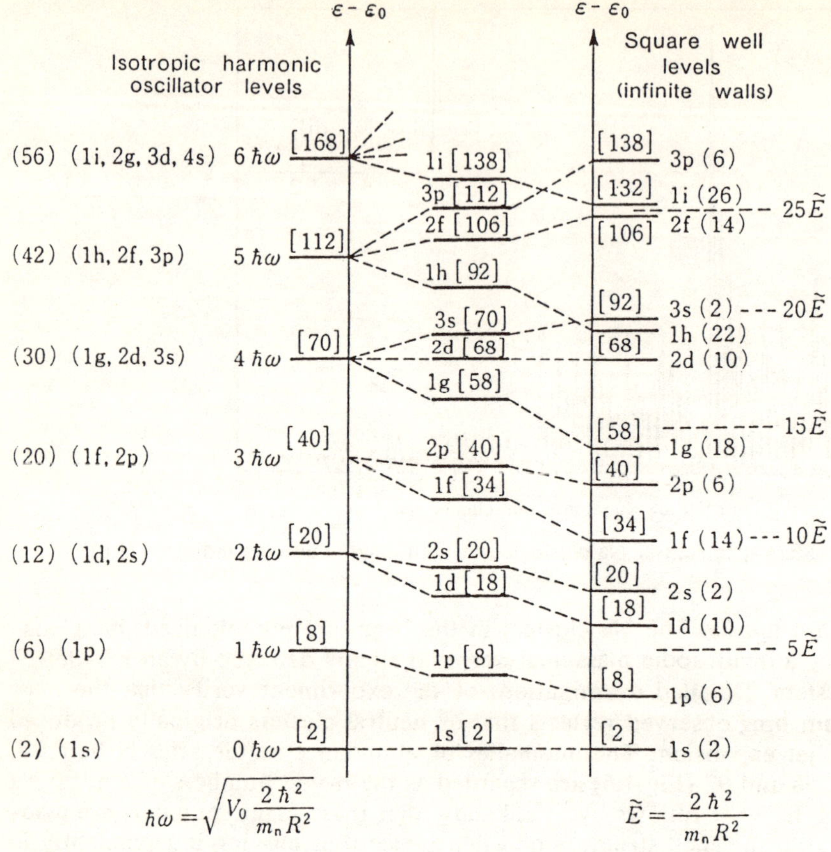

Fig.1.6. Energy level diagram of a single particle in effective potentials [1.9]

tive potential. The ion cores cannot have a shell structure as they are Bosons.

So far, it has been shown that full account of boundary conditions and the resulting *inhomogeneous* distributions of energy levels of valence electrons are important for metal microclusters. This is in contrast to the case of metal fine particles where the boundary problems are statistically treated and the resulting *homogeneous* distributions of the energy levels are utilized for deriving their properties at low temperatures. Such a situation of metal microclusters seems to show that the microclusters belong to the microscopic world like atoms and molecules, while the fine particles do to the macroscopic world. This is true in some aspects, but not so in every aspect. In Chap.2 we shall discuss that, at finite internal temperatures, microclusters may reveal the liquid phase as encountered in the macrosocpic world. In Chap.3, we shall discuss that collective excitations of metal microclusters may be described by using the classical or semi-classical liquid-drop model as in the case of heavy nuclei. The importance of boundary conditions or surface effects has also been revealed for non-metallic microclusters. For example, magic numbers have been observed in the mass spectrum of a Si cluster beam, as displayed in Fig.5.4, showing that Si_6 and Si_{10} clusters are relatively stable. As discussed in detail in Chap.5, the stable structures of these clusters calculated non-empirically are quite different from those expected from the sp^3 covalent bonding as found in the diamond structure. The difference comes from many dangling bonds associated with the surface atoms of microclusters, where the number of surface atoms is larger than that of the inside atoms: all the atoms are surface atoms in Si_6 and Si_{10}. This exmaple supports the classification of non-metallic microclusters according to the ratio of the number of inside atoms to that of surface atoms, as depicted in Fig.1.1.

1.2.3 Molecules

Aggregates of a few atoms are called molecules. Then, the following question arises: Is there any boundary between molecules and microclusters? If no distinct boundary exists, microclusters may be called large molecules. In what follows, we shall show that aggregates consisting of one kind (or a few kinds) of atoms have nearly degenerate low-lying metastable states whose number increases almost exponentially as the size N increases, and that the presence of a large number of these metastable states let even small aggregates of $N \sim 10$ show, at finite temperatures, physical properties often encountered in large systems: an example is the appearance of a liquid phase in small clusters as will be discussed in the next chapter. Because of this difference at finite temperatures between molecules and clusters of size $N \sim 10$, the lower boundary of microclusters is drawn at $N \sim 10$ in Fig.1.1.

The exponential increase of the number of metastable states, as the size N increases, may be seen in the calculation of local minima of the total energy of a cluster in the configuration space by using appropriate interaction potentials between two atoms. The number of local minima of N-atom clusters ($N = 6, 7, \ldots, 13$) obtained by using the Lennard-Jones potential,

$$v_{LJ}(r) = -2r^{-6} + r^{-12} , \tag{1.6}$$

are given in Table 1.1. Here, the constituent atoms are assumed to be indistinguishable. If the atomic configurations with the same spatial geometry but obtained by permutation or cooperative motion of atoms, as displayed in Fig.2.14 for the metastable state of the N = 6 cluster, are counted separately, the increase of the number of local minima begins much earlier, already at N = 6. Such a situation lets the N = 6 cluster show the liquid phase as

Table 1.1. The number of local minima of the total energy of a N-atom cluster, g(N), obtained by the use of the Lennard-Jones potential [1.10]

N	6	7	8	9	10	11	12	13
g(N)	2	4	8	18	57	145	366	988

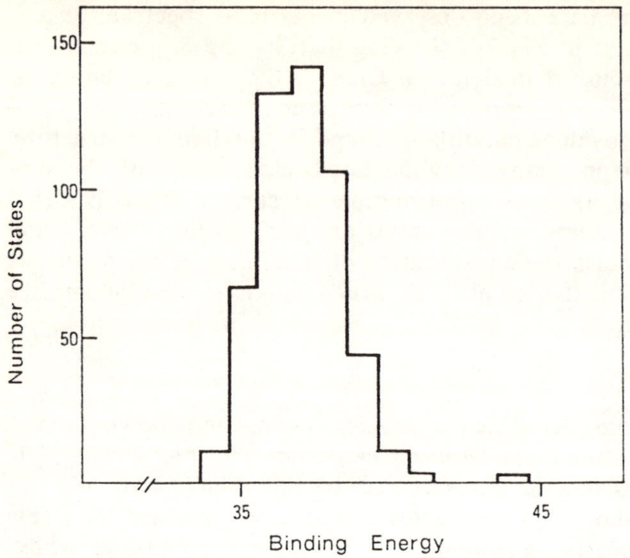

Fig.1.7. Distribution of stable and metastable states of the N = 13 cluster calculated by the Lennard-Jones potential. The energy scale is in units of the Lennard-Jones pair energy [1.10]

discussed in the next chapter. In this argument, the atomic configurations obtained by rotating the cluster as a whole are not counted separately.

Figure 1.7 shows the calculated distribution of the stable and metastable states of the N = 13 cluster. It is seen that many metastable states are distributed in a small range of energy separated from the stable ground state with a moderate gap. Generally, the magnitude of the gap depends upon the cluster size N and the interatomic potential. If it is comparable to thermal energies, we may expect thermal fluctuation of the atomic structure, as will be discussed in the next chapter. The presence of many metastable states in the narrow energy range also makes the excitation energy be dispersed into and stored in them for a relatively long time. Because of the finite number of freedom of the motion, however, there is a certain probability of the dispersed energy to be concentrated into a particular mode to induce fragmentation of a cluster. Such a process is somewhat similar to that found in the compound nucleus reaction.

2. Dynamics of Atomic Structure

Fluctuation of atomic structures of metal microclusters is observed by using an electron microscope of atomic resolution with a television camera. Such a structural fluctuation is one of the characteristics of microclusters. By using a computer, dynamical properties of small microclusters are studied for the size N = 13 bound by the Lennard-Jones interaction. The coexistence of solid-like and liquid-like phases is pointed out at the effective temperature between those for melting and freezing. On the other hand, for the transition-metal cluster of size N = 6, a fluctuating state accompanying no atomic diffusion is found just below the melting temperature, and a transition to permutation isomers accompanying atomic diffusion just above the melting temperature. Thermal properties of transition-metal microclusters are studied by using Monte-Carlo simulations in both cases of absence and presence of a magnetic interaction. It is observed that a sharp magnetic transition exists even in the small magnetic cluster of size N = 7.

2.1 Solid-like, Liquid-like, or Fluctuating?

2.1.1 Naive Questions

In the previous chapter, microclusters have been defined as aggregates of atoms, in which the number of surface atoms is larger than that of inside atoms. This induces our speculation that atomic diffusion may occur much more easily in microclusters than in bulk, as atoms feel less restriction at the surface as compared with in the inside. Then, the following naive questions would be raised: Are microclusters like solids where atoms are oscillating around their respective equilibrium position? Are microclusters like liquids where atoms are making diffusive motion? Or, are their atomic structures fluctuating between those of different solid phases?

As far as we know, no definite experimental determination has been succeeded of the atomic structures of free microclusters, although elaborate experiments of electron diffraction have been done on argon microclusters in a beam produced in a free jet expansion [2.1]. On the other hand, a temporal change in the atomic structure of metal microclusters adsorbed on a solid surface has been observed by using an electron microscope with atomic resolution. This kind of experiment will briefly be described in the next subsection.

2.1.2 Atomic Structure of Adsorbed Microclusters

The following experiment [2.2] has given a great impetus to the studies of dynamical properties of microclusters, irrespective of the results being related to intrinsic or extrnsic properties of microclusters. Gold clusters of 10÷100 Å radii are prepared in high vacuum on the surface of a Si substrate covered with SiO_2 film. The atomic structures of these clusters and their temporal change are observed by using an electron microscope with a resolution of 2.3 Å at 120 keV whose image in monitored by a television camera. The time resolution of the apparatus is 1/60 s.

When the cluster size is in the microcluster range, where the radius is less than 20 Å and the number of atoms contained is less than 10^3, we can notice that the clusters move on the surface as if they were alive. We can observe the fusion of two clusters when they are colliding. In addition to these movements, we can also see a temporal change of the atomic structure, as depicted in Fig.2.1. The pictures are selected from a VTR (Video Tape Recorder) over a 5-minute span. We note that the cluster consists of about 460 atoms. The structures in Figs.2.1c,e,f, and k are cuboctahedrons (Fig. 2.2) which are the local structures in a face-centered cubic crystal. The structures in Fig.2.1b and h are icosahedrons (Fig.2.3) never found as the local structures in any periodic crystal. Figure 2.1j displays a spherical cluster exhibiting no lattice fringe image. This might be a liquid droplet. Such a fluctuation of the atomic structure, as shown in Fig.2.1, cannot be observed when the cluster size is in the fine-particle range.

Carefully examining the dependence of the fluctuation upon the intensities of electron beams and the substrates, it is concluded that the fluctuation is caused by the charge fluctuation around and/or on the cluster. More recently, the dynamical properties of gold microclusters have also been studied by using high-resolution UHV electron microscopy with careful control of the temperature [2.4], concluding that the structural fluctuation of the microcluster is caused by the elevation of the internal temperatures. This conclusion means that the fluctuation is an intrinsic property of microclusters.

2.2 Coexistence of Solid-like and Liquid-like Phases

Computer simulations are performed on Ar_{13} clusters by using the method of Molecular Dynamics (MD) [2.5]. The MD calculation solves the Newtonian equations of motion in a given potential with appropriate initial conditions. In the simulations, a Lennard-Jones potential as given in (1.6) is assumed. The time step is 10^{-14}s. The initial conditions are determined to yield a given value of the total energy E_{tot}. The internal temperature T versus E_{tot} curve, called the caloric curve, is calculated, as depicted in Fig. 2.4. The definition of internal temperature T is

$$T = \frac{2}{3N-6}\langle K\rangle , \qquad (2.1)$$

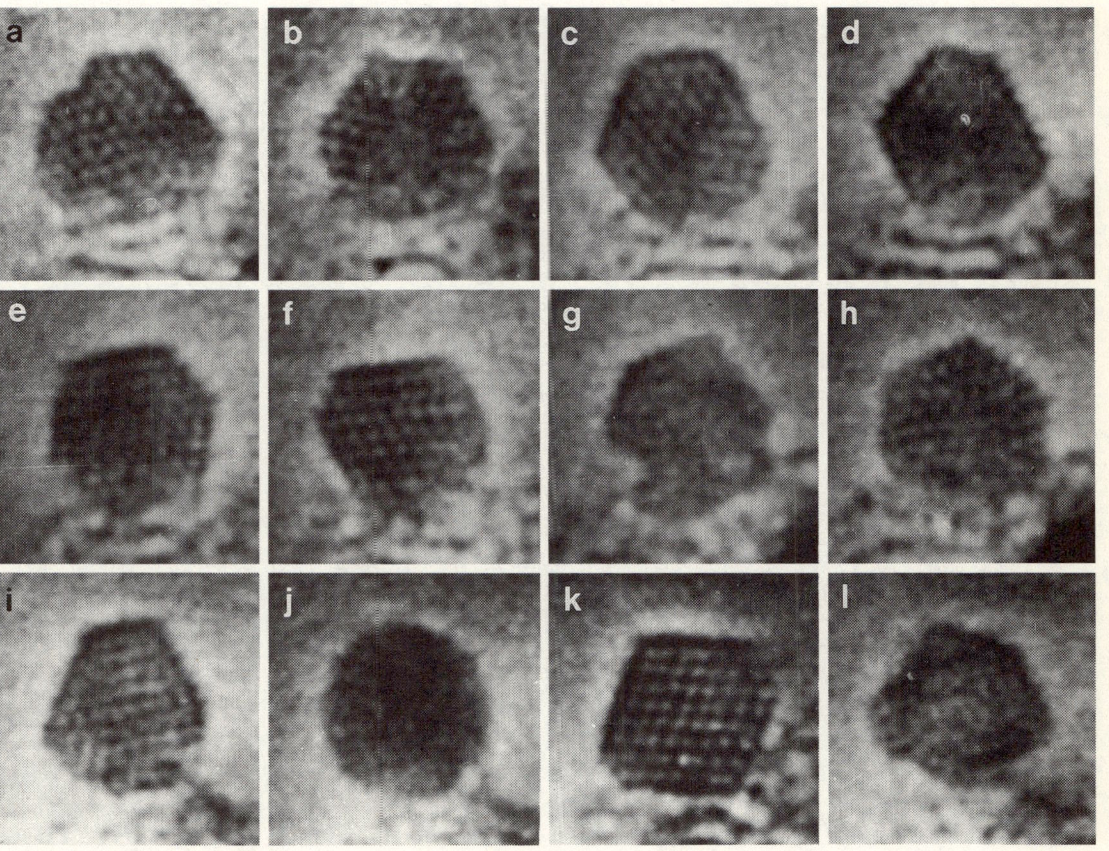

Fig.2.1. A series of the electron micrographs showing structural changes of a gold cluster consisting of about 460 atoms. For the details, see the text [2.2]

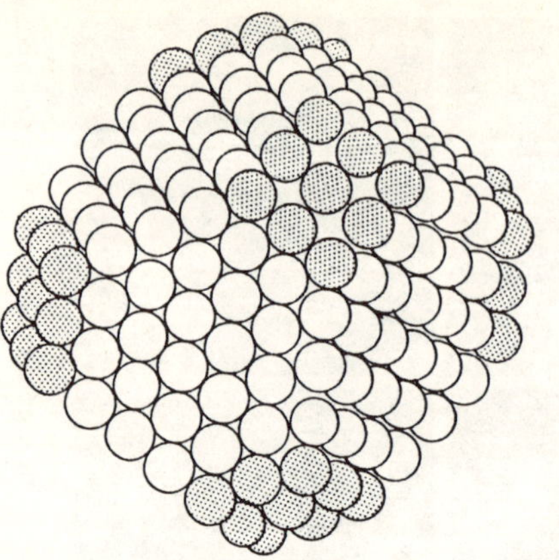

Fig.2.2. A cuboctahedral cluster consisting of 459 atoms [2.2]

Fig.2.3. An icosahedral cluster consisting of 561 atoms [2.3]

where K is the total kinetic energy, and $\langle \cdots \rangle$ means the time average. The factor (3N-6) results from the conservation of the total linear and angular momenta in the MD simulations. The time average is calculated in two ways: by the short-time average over 500 time steps corresponding to a few oscillations of the breathing vibrational mode, and by the long-time average over $5 \cdot 10^4 - 1 \cdot 10^6$ time steps.

The caloric curve in Fig.2.4 may be divided into three parts; (i) $E_{tot} < E_f$, (ii) $E_f < E_{tot} < E_m$, and (iii) $E_m < E_{tot}$. In parts (i) and (iii), the long- and short-time averages practically coincide. In part (ii), the short-time averages generate two branches of the caloric curve and the long-time averages an averaged curve of these two branches. In what follows, it is discussed that (i) corresponds to the solid-like phase, (iii) to the liquid-like phase, and (ii) to the coexistence state of the solid-like and liquid-like

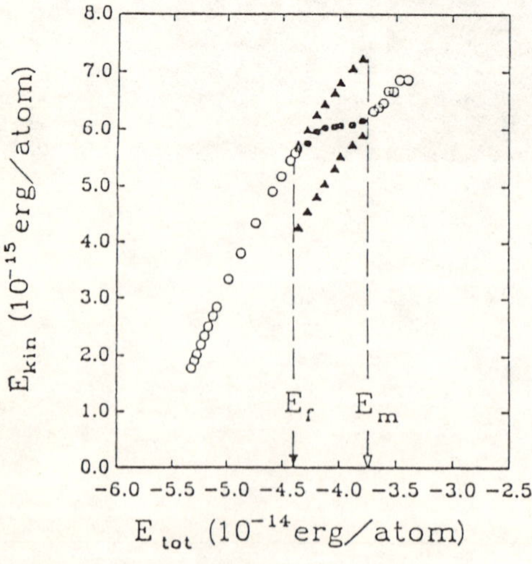

Fig.2.4. The caloric curve of the Ar_{13} cluster. Points ○ and • correspond to the long-time averages while ▲ to the short-time averages [2.4]

phases. The internal temperatures at E_f and E_m are called, respectively, the freezing and melting temperatures, T_f and T_m.

In order to examine characteristic properties of the states in various parts of the caloric curve, let us calculate such quantities as the root-mean-square (rms) bond-length fluctuation δ and the mean-square displacement $\langle r^2(t)\rangle$, which are defined as follows

$$\delta = \frac{2}{N(N-1)} \sum_{i<j} \frac{(\langle r_{ij}^2\rangle - \langle r_{ij}\rangle^2)^{1/2}}{\langle r_{ij}\rangle} \tag{2.2}$$

where $r_{ij} = |\mathbf{r}_i - \mathbf{r}_j|$ is the distance between the ith and jth atoms, and $\langle\cdots\rangle$ is the long-time average

$$\langle r^2(t)\rangle = \frac{1}{Nn_t} \sum_{j=t}^{n_t} \sum_{i=1}^{N} [\mathbf{r}_i(t_{0j}+t) - \mathbf{r}_i(t_{0j})]^2 \ , \tag{2.3}$$

n_t being the number of different time origins t_{0j}.

The calculated bond-length fluctuation as a function of the internal temperatures is shown in Fig.2.5. The abrupt change of δ at T = 34 K corresponding to the region of $E_{tot} \sim E_f$ and E_m is remarkable. The change occurs when δ approaches the value of 0.1. This reminds us of the Lindemann criterion, stating that bulk melting begins when the bond fluctuation just exceeds 10%. At this stage, we tentatively assign parts (i) and (iii) of the caloric curve to the solid-like and liquid-like phases, respectively.

The calculated mean-square displacement curves as functions of time at $E_{tot} = -4.72 \cdot 10^{-14}$ erg/atom < E_f (lower curve) and $E_{tot} = -3.61 \cdot 10^{-14}$ erg/atom > E_m (upper curve) are exhibited in Fig.2.6. Since the slopes of the long-time part of these curves are related to the diffusion coefficient D as

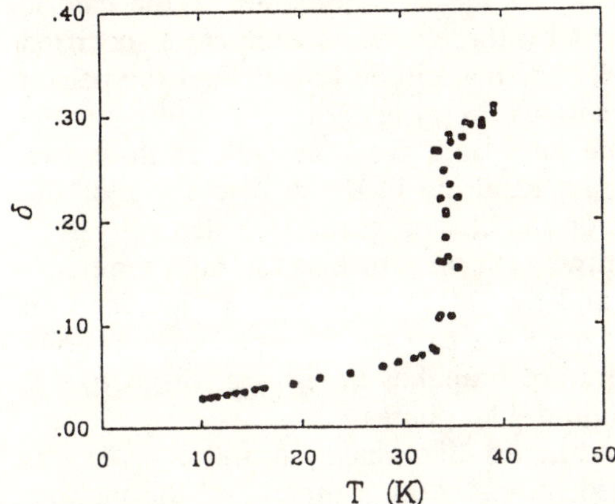

Fig.2.5. The calculated bond-length fluctuation as a function of the internal temperature T [2.4]

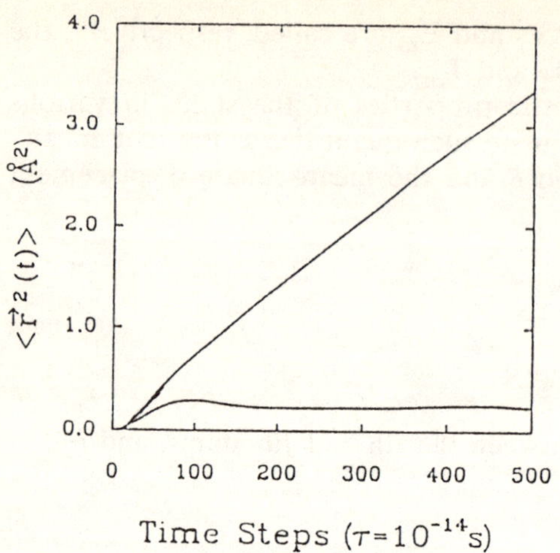

Fig.2.6. The mean-square displacement versus time at $E_{tot} < E_f$ (lower curve) and $E_{tot} > E_m$ (upper curve) [2.4]

$$D = \frac{1}{6}\frac{d\langle r^2(t)\rangle}{dt}, \tag{2.4}$$

the lower and the upper curves indicate, respectively, the solid-like and liquid-like phases in agreement with the tentative assignment derived from the temperature dependence of the bond length fluctuation shown in Fig. 2.5. It should be noted here that the times to define D may extend only to the period required for an atom to migrate across the diameter of the cluster.

Now we are at the position to examine (ii) of the caloric curve in detail. Figure 2.4 shows that the two branches of the short-time averages in this portion look like continuations of the solid-like and the liquid-like curves in (i) and (iii), respectively. This observation makes us believe that the state of the cluster in this temperature range (or this total energy range) would be alternating between the solid-like and the liquid-like, staying in each one of them for short-time periods. This supposition is confirmed by the results in Fig.2.7, which reveals fluctuation of the internal temperatures between the two values corresponding to the two branches of the caloric curve in (ii). The time intervals spent by the cluster in each state specified by the temperature T^l or T^h are orders of magnitude longer than the period of the characteristic vibrational motions in either the solid-like or the liquid-like state. The fraction of the total time spent in each of these two states is a function of the total energy, resulting in the caloric curve of the long-time average in (ii). Snapshots of the cluster in the two states given in Fig.2.7 indicate two different structures. The structure at the high temperature T^h is a vibrating icosahedron and that at T^l looks like an icosahedron with one atom plucked out and put onto the surface. These snapshots confirm that the high- and low-temperature branches in (ii) are, respectively, continuations of the solid-like and liquid-like phases.

The coexistence of solid-like and liquid-like phases in microclusters, as described here, was already discussed in a general argument of the melting

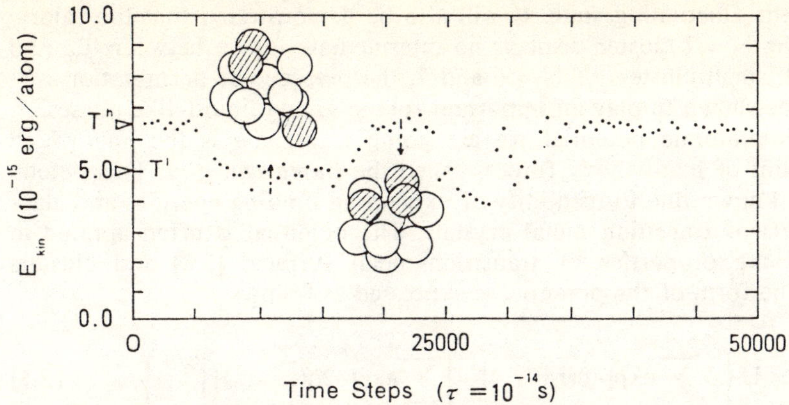

Fig.2.7. The time-dependence of the internal temperature T for E_{tot} = $-4.16 \cdot 10^{-14}$ erg/atom. Snapshots of the cluster at T^l and T^h are also shown [2.4]

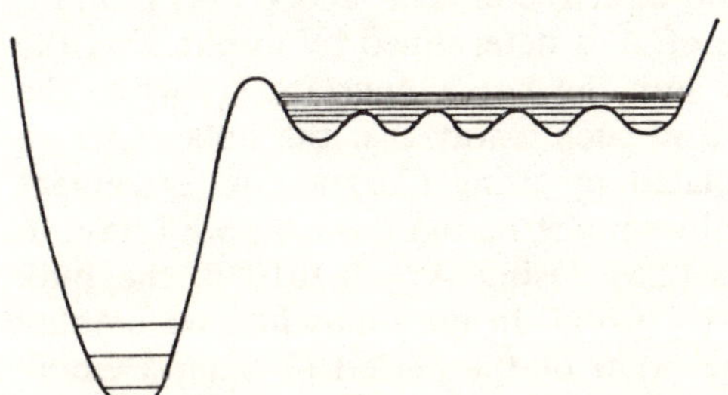

Fig.2.8. Potential with a deep well giving a solid-like state and a broad region of higher energy with shallow minima giving a liquid-like state [2.6]

of finite clusters [2.6] before performing the computer simulations of molecular dynamics of Ar_{13} clusters. In this argument, a parameter playing a role analogous to the order parameter is introduced. The parameter is so defined to measure the rigidity of the system. The argument seems to be particularly suitable for the system having the form of a potential (Fig.2.8). In the following sections it will be shown that behaviors of the transition from the solid-like to the liquid-like state are more complicated and depend upon the form of the potential and, consequently, upon the size N of the clusters when N is small.

2.3 Fluctuating States and Permutation Isomers

Detailed studies of the dynamics of atomic structures are done in small clusters of N = 6 and 7 with a different interatomic potential [2.7]. It will be demonstrated that a caloric curve similar to that of Ar_{13} is obtained for the N = 6 cluster, but in the temperature range between T_m and T_f only continual transitions between the ground and metastable states are found. The state of such continual transitions, called a fluctuating state, is entirely different from the coexistence state in such a point that no atomic diffusion occurs in the fluctuating state. It will also be demonstrated that the caloric curve of the N = 7 cluster displays no intermediate region between T_m and T_f. In such small clusters of N = 6 and 7, the presence of permutation isomers will be shown to play an important role in giving liquid-like phases.

The interatomic potential we are going to assume is the one which takes account of many-body forces, called the *Gupta potential*. This potential is well known due to its ability of explaining binding energies and elastic constants of transition metal crystals. The potential is often applied to studies of the properties of transition-metal surfaces [2.8] and clusters [2.9, 10]. The form of the potential is expressed as follows

$$V(r_i) = U\left\{A \sum_j \exp[-p(r_{ij}-r_0)] - \left[\sum_j \exp[-2q(r_{ij}-r_0)]\right]^{1/2}\right\}, \tag{2.5}$$

where r_0 is the lattice constant of the bulk fcc crystal; and U, A, p and q are adjustable parameters. The value of A is determined by minimizing the cohesive energy of the bulk crystal with the lattice constant r_0, while the values of U, p and q are determined in such a way that the bulk cohesive energy and the bulk modulus calculated by using (2.5) are in agreement with the experimental values. The values $p = 9/r_0$ and $q = 3/r_0$ are found to be appropriate for transition metals [2.8]. Using A = 0.101035, the bulk cohesive energy is given by $E_{bulk} = 1.17674U$. In the following, we employ r_0, E_{bulk}, and $r_0(m/E_{bulk})^{-1/2} \sim$ (the order of the period of atomic vibrations, 10^{-13}s) as the units of distance, energy and time, respectively. The use of these units makes the following arguments applicable to clusters of any transition-metal atoms. In (2.5) a many-bodies interaction is included through the square-root of the second sum. This kind of interaction increases the strength of a bond when the number of bonds stretching from an atom is decreased.

2.3.1 The Case of N = 6

As also found in the N = 6 cluster with a Lennard-Jones potential, two minima are seen in the potential-energy surface calculated with a Gupta potential; the global minimum giving an OCTahedral (OCT) structure and a local one a TriPyramid (TP), as depicted in Fig.2.9. The MD calculation is carried out with the initial conditions for the atomic coordinates to form OCT and for the atomic velocities to be randomly distributed with the translational and rotational degrees of freedom frozen.

The caloric curve and the specific heat per atom as a function of the total energy calculated by the MD method are exhibited in Figs.2.10 and 11, respectively. The internal temperature T is given by (2.1) and the specific heat by [2.7].

$$C = \left[N - N\left(1 - \frac{2}{3N-6}\right)\langle K\rangle\langle K^{-1}\rangle\right]^{-1}, \tag{2.6}$$

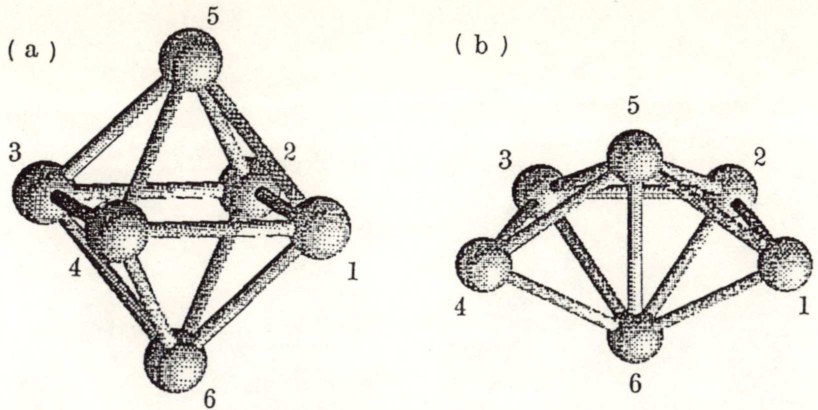

Fig.2.9. (a) The stable structure of octahedron (OCT) and (b) the metastable structure of tripyramid (TP) of the N = 6 cluster

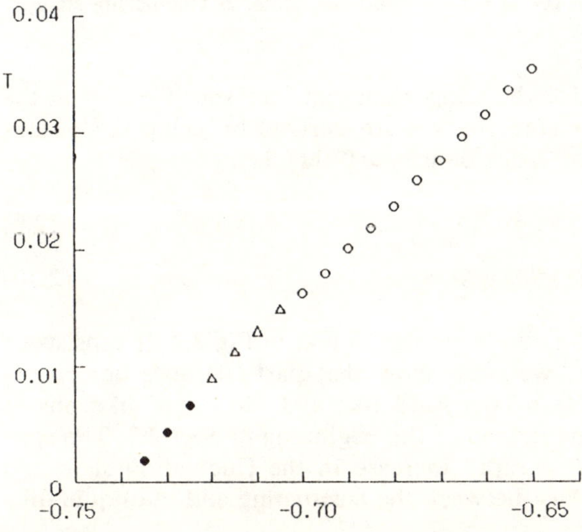

Fig.2.10. Caloric curve for N = 6: • solid-like state, Δ fluctuating state, o liquid-like state [2.7]

which is related to the fluctuation of the kinetic energy as

$$\langle K^{-1}\rangle = \frac{1}{\langle K\rangle}\langle\frac{1}{1+\delta K}\rangle = \frac{1}{\langle K\rangle}[1 + \langle(\delta K)^2\rangle + \langle(\delta K)^4\rangle + \ldots] , \tag{2.7}$$

where

$$\delta K = \frac{K - \langle K\rangle}{\langle K\rangle} . \tag{2.8}$$

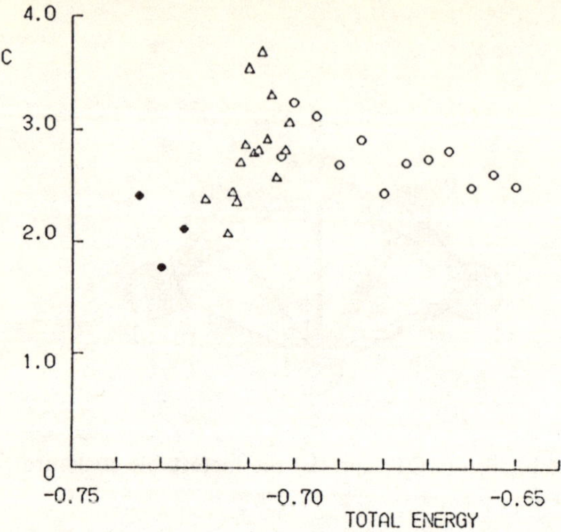

Fig.2.11. Specific heat per atom for N = 6: • solid-like state, Δ fluctuating state, ○ liquid-like state [2.7]

The expressions (2.1 and 6) for the temperature and the specific heat in the microcanonical ensemble for free clusters are derived by using the phase-space volume Ω and the phase-space density ω [2.11], i.e.,

$$T = \Omega/\omega \tag{2.9}$$

$$C = (N - NT\omega^{-1}\partial^2\Omega/\partial E^2)^{-1} . \tag{2.10}$$

The caloric curve in Fig.2.10 is quite similar to that in Fig.2.4. It consists of three parts. In what follows, we shall show that part (ii) does not correspond to the coexistence state of the solid-like and the liquid-like phases but to the fluctuating state mentioned at the beginning of Sect.2.3. The specific heat in Fig.2.11 reveals a rapid increase in the fluctuating state and has a peak around the boundary between the fluctuating and the liquid-like states.

In order to see a temporal change of the atomic structure in a simple way, we ask the computer to calculate the distance index d defined as

$$d = \tfrac{1}{2}\|\mathbf{A} - \mathbf{A}_{OCT}\|^2 , \tag{2.11}$$

where $\mathbf{A}$ is the adjacency matrix whose elements A_{ij} are given as

$$A_{ij} = \begin{cases} 1 , & \text{for } |\mathbf{r}_i - \mathbf{r}_j| < r_n \\ 0 , & \text{otherwise .} \end{cases} \tag{2.12}$$

In our problem, $r_n = 1.2$ is found to be more appropriate than $r_n = 1.0$ empirically. Anyhow, matrix A represents how the constituent atoms are connected by relatively strong bonds. In (2.11) the norm of a matrix is defined by

$$\|\mathbf{A}\| = \Big(\sum_i \sum_j A_{ij}{}^2\Big)^{1/2} . \tag{2.13}$$

Matrix $\mathbf{A}_{OCT}$ is the adjacency matrix $\mathbf{A}$ for the OCT structure. By using the numbering of atoms as in Fig.2.9a and assuming the lengths of the bonds indicated in the figure to be less than 1.2, matrix $\mathbf{A}_{OCT}$ is easily obtained as

$$\mathbf{A}_{OCT} = \begin{vmatrix} 1 & 1 & 0 & 1 & 1 & 1 \\ 1 & 1 & 1 & 0 & 1 & 1 \\ 0 & 1 & 1 & 1 & 1 & 1 \\ 1 & 0 & 1 & 1 & 1 & 1 \\ 1 & 1 & 1 & 1 & 1 & 0 \\ 1 & 1 & 1 & 1 & 0 & 1 \end{vmatrix} . \tag{2.14}$$

For the purpose of later use, we also give here the adjacency matrix $\mathbf{A}$ for the TP structure, $\mathbf{A}_{TP}$. As seen in Fig.2.9b, the TP structure is obtained from the OCT structure by cutting off the bond between atoms 1 and 4, and producing a bond between atoms 5 and 6. By using Fig.2.9b, it is easy to derive

$$\mathbf{A}_{TP} = \begin{vmatrix} 1 & 1 & 0 & 0 & 1 & 1 \\ 1 & 1 & 1 & 0 & 1 & 1 \\ 0 & 1 & 1 & 1 & 1 & 1 \\ 0 & 0 & 1 & 1 & 1 & 1 \\ 1 & 1 & 1 & 1 & 1 & 1 \\ 1 & 1 & 1 & 1 & 1 & 1 \end{vmatrix} . \tag{2.15}$$

The distance index d calculated as a function of time for the total energies in three portions of the caloric curve of Fig.2.10 are given in Fig. 2.12a-c. In Fig.2.12a, where the total energy is assumed to be in the energy range of part (i) of Fig.2.10 indicated by black circles, index d is fluctuating between the values of 0 and 1. The state of d = 1 corresponds to the one in which one bond in the OCT structure is broken or an additional bond is formed. This state is realized by increasing vibrational amplitudes of specific atoms. Thus, we see that part (i) corresponds to the solid-like phase, in which the constituent atoms are vibrating around the fixed stable points.

On the other hand, in Fig.2.12b, index d sometimes takes the value of 2 besides the values of 0 and 1. Detailed examination of the d = 2 configu-

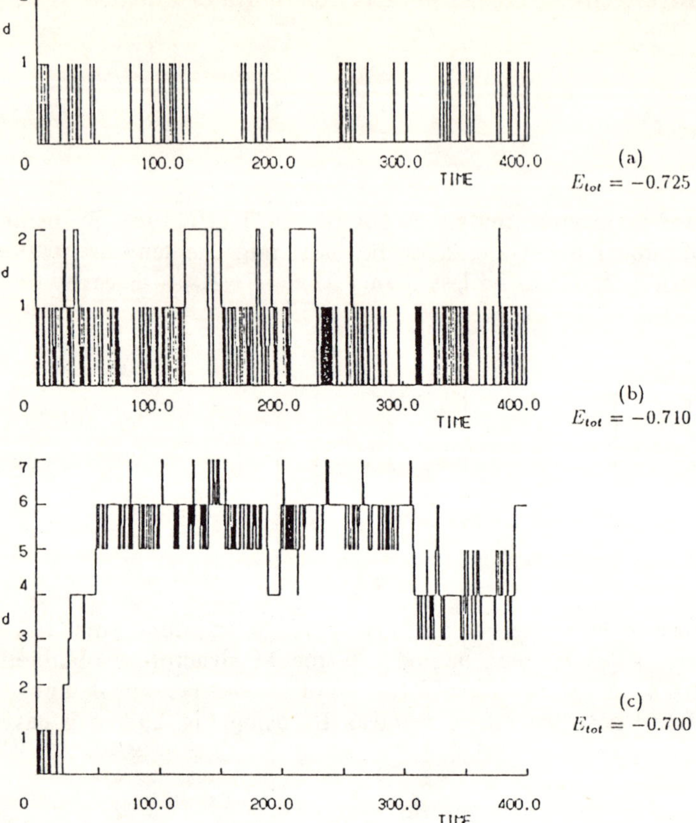

Fig.2.12. Time evolution of distance index d in **(a)** the solid-like, **(b)** fluctuating, and **(c)** liquid-like states [2.7]

ration appearing here indicates that it is the TP. Note that, by using (2.14 and 15), one obtains

$$\tfrac{1}{2}||\mathbf{A}_{TP} - \mathbf{A}_{OCT}||^2 = 2 \,. \tag{2.16}$$

Thus we see that the system in the energy range of part (ii) of Fig.2.10 is fluctuating between the ground and the metastable states. We call such a state in part (ii) a *fluctuating state*. As already mentioned at the beginning of Sect.2.3, the fluctuating state is entirely different from the coexistence state introduced in Sect.2.2 in such a way that no atomic diffusion occurs in the fluctuating state.

The time evolution of index d in the energy range of part (iii) is quite different from those in parts (i) and (ii), as shown in Fig.2.12c. It takes the value of 6 after some passage of time and fluctuates around this value with the amplitudes ±1. Then it goes to the value of 4 and fluctuates again around this value with the same amplitudes. Detailed examination of the d =

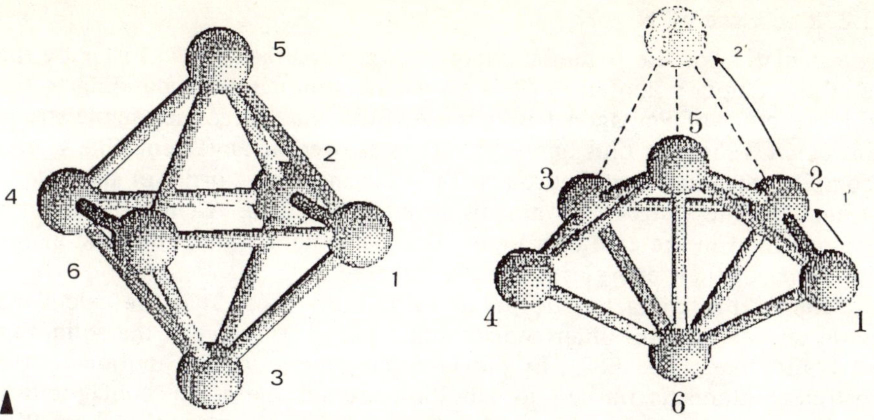

▲

Fig.2.13. The permutation isomer OCT′ obtained by two successive permutations, (3↔4) and then (3↔6), from the OCT in Fig.2.9a

Fig.2.14. Permutation isomer (1′2′3456), TP′, obtained by the cooperative motion of two atoms, 1 and 2.

6 configuration appearing here reveals that it is OCT′ obtained by two successive permutations, (3↔4) and then (3↔6), from the OCT in Fig.2.9a, as shown in Fig.2.13. One can confirm the relation

$$\tfrac{1}{2}\|\mathbf{A}_{\mathrm{OCT}'} - \mathbf{A}_{\mathrm{OCT}}\|^2 = 6\,. \tag{2.17}$$

We call OCT′ a permutation isomer of OCT. Cutting off the bond between atoms 2 and 4, and letting atoms 3 and 5 move closer to form a bond between them, we obtain TP′, a permutation isomer of TP, as indicated in Fig.2.14. In the figure, this TP′ is shown to be also obtained from TP by the cooperative motion of two atoms, 1 and 2. Detailed examination of the d = 4 configuration appearing in Fig.2.12c exhibits that it is TP′, as shown in Fig.2.14. One can confirm the relation

$$\tfrac{1}{2}\|\mathbf{A}_{\mathrm{TP}'} - \mathbf{A}_{\mathrm{OCT}}\|^2 = 4\,. \tag{2.18}$$

Note that the transition TP→TP′ is not a rotational motion: in our calculation the translational and rotational degrees of freedom are frozen. Summarizing these examinations, we may conclude that in the low-temperature region of the liquid-like state of such a small cluster of N = 6, continual transitions among the permutation isomers of the stable and the metastable configurations, accompanying cooperative motion of atoms, characterize dynamical behaviors of the system. We remark here that the number of permutation isomers is relatively large, even in such a small cluster of N = 6, and this would be the reason why the liquid-like state, as found in large systems, appears in such a small system, too.

2.3.2 The Case of N = 7

The calculation of the potential energy surface for the N = 7 cluster by the use of the Gupta potential in (2.5) gives four minima corresponding to the stable structure of Pentagonal BiPyramid (PBP), and three metastable structures of OCTahedron plus one (OCT+1), a skewed arrangement (Skew) and Incomplete Stellated Tetrahedron (IST) in the increasing order of the potential energy. These structures are displayed in Fig.2.15. All these structures are also found in the case of a Lennard-Jones potential, although the increasing order of their energies is slightly different.

One of the characteristic points of this system is that the calculated caloric curve lacks the intermediate state, part (ii), between the solid-like and liquid-like states [2.7]. In the low-energy region, the motion of the constituent atoms is limited to vibration around the stable configuration PBP. When the total energy exceeds a critical value, the transition from PBP to (OCT+1) occurs. At this critical temperature we see an abrupt increase of the fluctuation of bond length, as also found in the N = 6 and 13 clusters. The state above this critical temperature may be considered to be the liquid-like state. The reason why no intermediate state appears is that the continual transition between PBP and (OCT+1) induces diffusive motion of atoms, as seen as follows: The first cycle of the transition may be achieved by shifting atom 1 in Fig.2.15a upward, making a bond between atoms 2 and 5, and cutting the bond between atoms 6 and 7. Once the (OCT+1) structure is formed, we see that three bonds between atoms 2 and 6, between 2 and 5, and between 5 and 6 are equivalent. (Fig.2.15b). Then, the

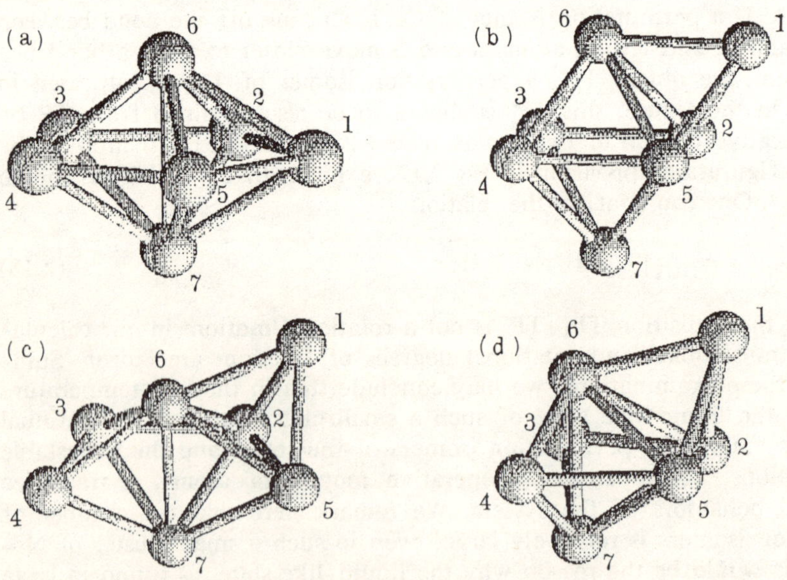

Fig.2.15a-d. The stable and metastable structures of the N = 7 cluster: **(a)** pentagonal bipyramid (PBP), **(b)** octaheron plus one (OCT+1), **(c)** skewed arrangement (Skew), and **(d)** incomplete stellated tetrahedron (IST)

following three equivalent motions coming back to the PBP structure are conceivable; (1) just the reversed motion, (2) to cut the 5-6 bond forming a pentagon 1-5-7-3-6, and to make a bond 2-4, (3) to cut the 2-6 bond forming a pentagon 1-2-7-4-6, and to make a bond 3-5. All these motions reconstruct the same PBP but different permutation isomers. Since transitions between different permutation isomers accompany diffusive motion of atoms, we see that the state of going back and forth between PBP and (OCT+1) is liquid-like, but not the fluctuating state as found in the N = 6 cluster.

When the total energy is further increased in the liquid-like state, transitions such as (OCT+1) → Skew, (OCT+1) → IST, Skew → IST, PBP → Skew, PBP → IST and the reversed ones are induced. All these transitions accompany diffusive motion of the atom. Some of them are illustrated in Fig.2.16.

The calculated diffusion constant D as a function of the total energy by the use of (2.3 and 4) is plotted in Fig.2.17. The figure shows that D is almost vanishing in the solid-like state and it increases gradually in the liquid-like state as the total energy increases. This behavior may be understood as follows: In the liquid-like state, the constituent atoms make diffusive motion vibrating for some time around the metastable configurations with a residence time much shorter than that in the solid-like state. The residence time becomes shorter as the total energy increases.

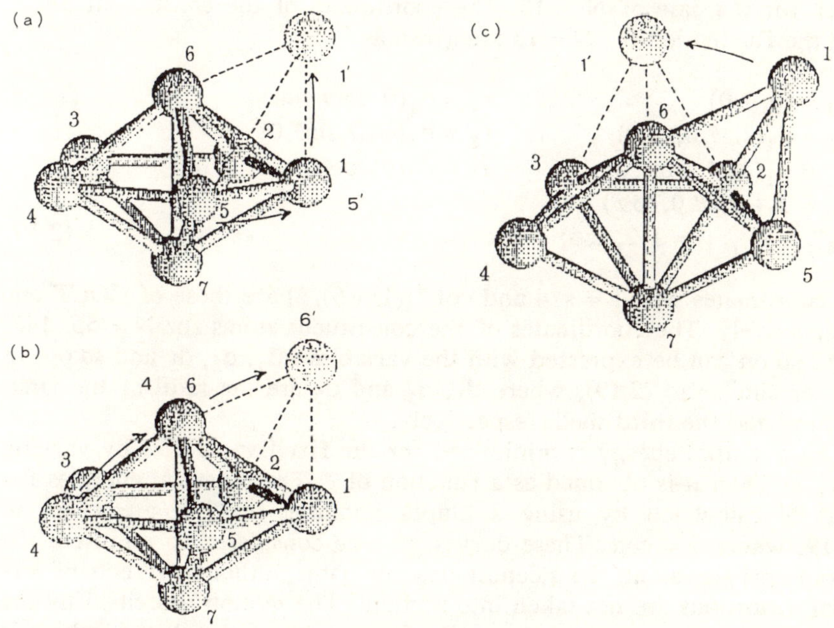

Fig.2.16a-c. Atomic motions in transitions: (a) PBP → Skew, (b) PBP → IST, (c) Skew → IST.

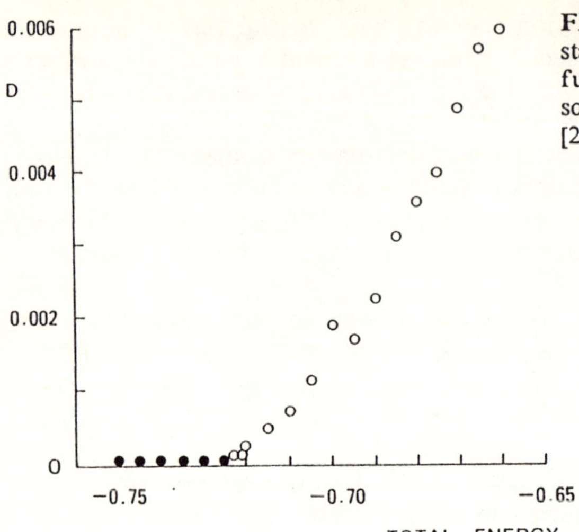

Fig.2.17. Calculated diffusion constant D for the N = 7 cluster as a function of the total energy: • solid-like state, o liquid-like state [2.7]

2.3.3 Fluctuation in Large Clusters

We discuss here the fluctuation between ICosahedron (IC) and CubOCTahedron (COCT) in large clusters of N = 13, 55, 147, 309 and so on. Such a fluctuation may be described by using a single parameter θ, as exhibited in Fig.2.18 for the case of N = 13. The coordinates of the constituent atoms during the fluctuation for N = 13 are given as

$$
\begin{aligned}
&\mathbf{r}_0 = (0,0,0) && \mathbf{r}_1 = d_1(0,\cos\theta,\sin\theta)\\
&\mathbf{r}_2 = d_1(\sin\theta,0,\cos\theta) && \mathbf{r}_3 = d_1(\cos\theta,\sin\theta,0)\\
&\mathbf{r}_4 = d_1(\cos\theta,-\sin\theta,0) && \mathbf{r}_5 = d_1(0,\cos\theta,-\sin\theta)\\
&\mathbf{r}_6 = d_1(-\sin\theta,0,\cos\theta) &&\\
&\mathbf{r}_{6+i} = \mathbf{r}_{7-i}\quad (i = 1,2,\dots,6)\ . &&
\end{aligned}
\tag{2.19}
$$

These coordinates with $\theta = \pi/4$ and $\cot^{-1}[(1+\sqrt{5})/2]$ are those of COCT and IC, respectively. The coordinates of the constituent atoms for N = 55, 147, 309 and so on can be expressed with the variables θ, d_1, d_2, d_3 and so on, in a manner similar to (2.19), where d_1, d_2 and d_3 are the radii of the first, the second and the third shell, respectively.

The potential energy is minimized for the fixed value of θ by varying d_1, d_2, ... Then it is obtained as a function of θ. The potential energies for various N calculated by using a Gupta potential in (2.5) are given in Fig.2.19, where $\sigma = \cot\theta$. These curves may be considered to represent the potential energies along the steepest-descent paths, although effects of additional distortions are not taken into account. The motion described by the parameter σ may be regarded as a collective motion of the cluster, which is responsible for the fluctuating states.

Generally speaking, the probability of activating a specific collective mode decreases as N increases. This is because the total number of the

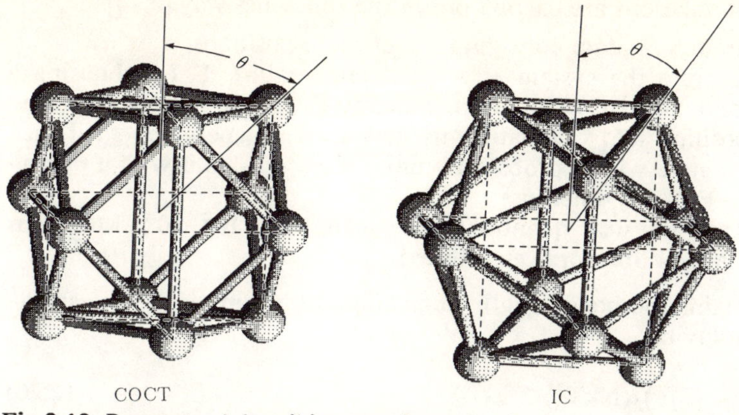

Fig.2.18. Parameter θ describing transformation between a cuboctahedron (COCT) and an icosahedron (IC)

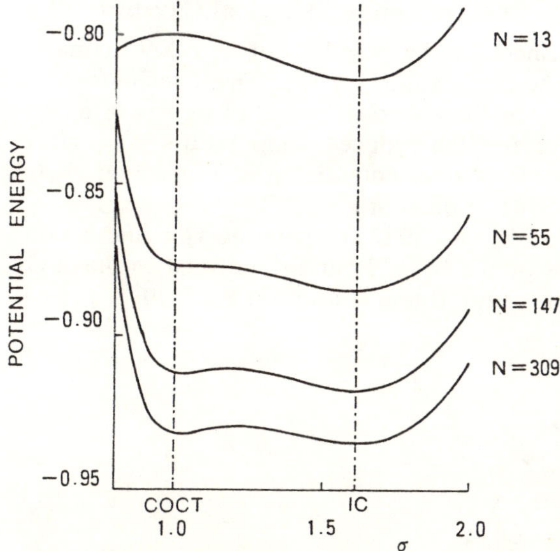

Fig.2.19. Potential energies of icosahedrons as a function of $\sigma = \cot\theta$ [2.7]

modes of cluster motion increases as N increases. Therefore, one may expect narrowing of the intermediate region of the fluctuating state in the caloric curve in large clusters.

2.4 Monte-Carlo Simulations

In Sects.2.2 and 3, we treated an isolated cluster, a microcanonical ensemble, by using the Molecular Dynamics (MD) simulations [2.5]. In this section we treat a cluster in contact with a heat reservoir, a canonical ensemble, by using the Monte-Carlo (MC) simulations [2.12, 13]

The MC simulations are carried out in the following way [2.14]:

1) to generate a random configuration of atom positions;

2) to equilibrate the system at a fixed temperature T, (i) choosing a random atom and calculating its internal energy E_1, (ii) moving it to a random nearby position and calculating its new internal energy E_2, (iii) accepting the new state with a probability unity if $E_2 < E_1$, but with a probability $\exp[-(E_2-E_1)/kT]$ otherwise;

3) to average physical quantities over many, say 10^5-10^6, MC steps after the thermal equilibrium is established.

Finally one obtains the bond-length fluctuation δ by using (2.2), the specific heat by employing

$$C = (\langle E^2\rangle - \langle E\rangle^2)/(NkT^2)\,, \tag{2.20}$$

and so on. In (2.20), E is the internal energy per atom.

2.4.1 Lennard-Jones Clusters Constrained to Spherical Cavities

The MC simulations are performed for the N = 13, 201 and 209 Lennard-Jones clusters confined to spherical cavities [2.15]. The confinement is achieved by placing the clusters in the spherical potential well with an infinite wall. In what follows we use the reduced temperature $T^* = kT/\epsilon$, where ϵ is the depth of the Lennard-Jones potential at its minimum: $\nu_{LJ}(r)$ in (1.6) is given by the energy scale in units of ϵ.

The specific heat calculated by use of (2.20) from energy fluctuations obtained in the MC simulations of the N = 13 cluster confined to the cavity, is shown as a function of the reduced temperature in Fig.2.20. A peak is

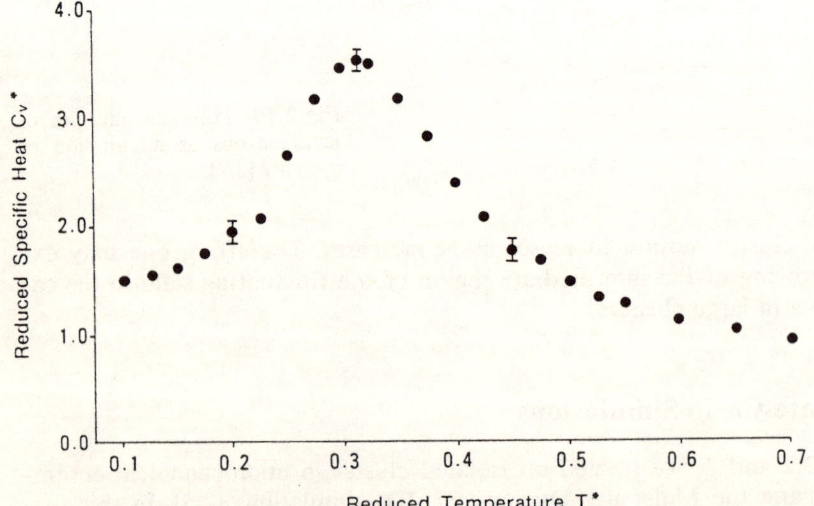

Fig.2.20. The specific heat for N = 13 calculated by MC simulations as a function of reduced temperature T^* [2.15]

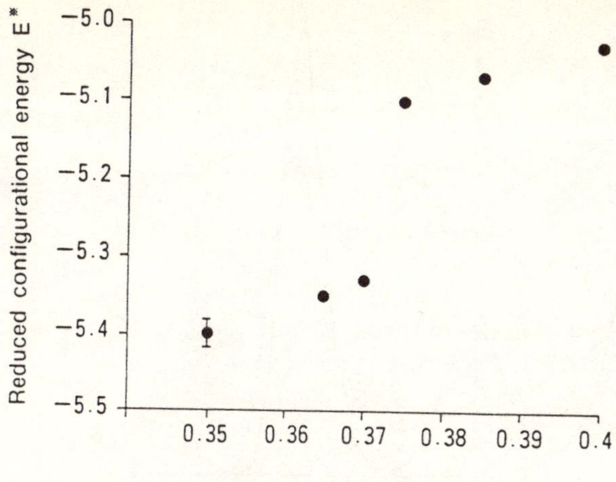

Fig.2.21. The caloric E^* vs. T^* curve for the N = 201 cluster. The reduced internal energy E^* is given by $E^* = E/N\epsilon$, E being the internal energy [2.15]

seen at $T^* = 0.31$. Inspection of the cluster structure at temperatures on either side of the peak indicates that the average structure is solid-like on the left of the peak and it is liquid-like on the right. The temperature giving the peak may be taken as the melting temperature T_m^*, although the identification can only be approximate. This melting temperature should be compared with the bulk triple-point temperature for the Lennard-Jones system, $T_{TP}^* = 0.68$ [2.16].

The calculated reduced internal energy E^* for N = 201 is plotted as a function of T^* in Fig.2.21. The melting transition is very sharp as found in the bulk, but the melting temperature determined from this curve is $T_m^* = 0.37$, which is still very low. A similar curve is also obtained for the N = 209 cluster and the melting temperature is found to be $T_m^* = 0.38$.

The changes in the three-body structure in the core of the N = 201 cluster on freezing are shown in Fig.2.22, where the function $f(\cos\theta)$, the fraction of interior angles of triplets of atoms with pair separations up to 1.35, is plotted against $\cos\theta$ for $T^* = 0.370$ and 0.375, respectively, corresponding to the temperatures after and before the freezing transition: here the pair separation is measured in units of the equilibrium distance of the Lennard-Jones potential. This function $f(\cos\theta)$ can be used to discriminate between different solid structures at low temperatures. For example, the perfect face-centered cubic (fcc) lattice has peaks of $f(\cos\theta)$ at $\cos\theta = -1$, ±0.5, 0.0, and the hexagonal closed packed (hcp) lattice at -1, -0.83, ±0.5, -0.33, 0.0, while the ICosahedral structure (IC) near -1, ±0.5. In Fig.2.22, the dashed curve corresponding to the liquid-like state has broad peaks at -1, ±0.5, indicating that the state is fluctuating around the IC ordering. The full curve for the solid-like state has sharp peaks at -1, -0.85, ±0.5, 0.0, indicating that the state would have the hcp ordering, although a peak at -0.33 is missing.

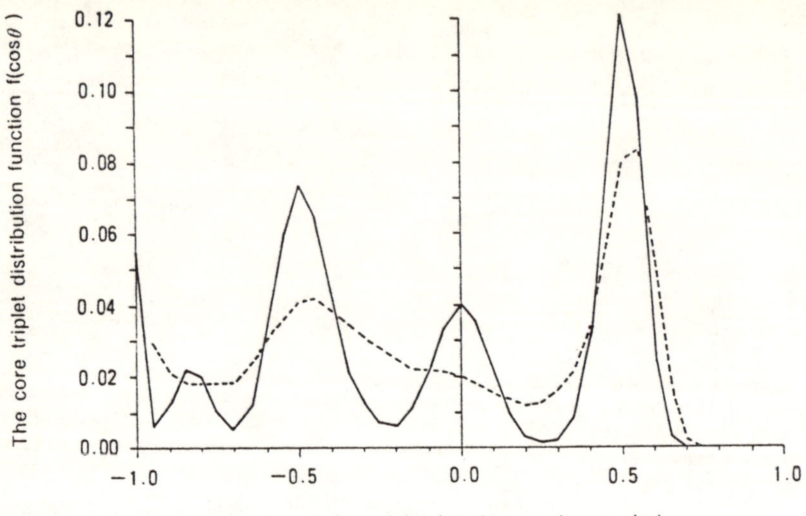

Fig.2.22. The distribution of interior angles θ of triplets of atoms in the core of the N = 201 cluster before ($T^* = 0.375$, dashed) and after the freezing transition ($T^* = 0.370$, full curve). Only interior angles formed at an atom whose center is within an atomic diameter of the center of the cavity are investigated and all triplets considered have sides of length smaller than or equial to 1.35 [2.15]

2.4.2 Transition-Metal Clusters of N = 7–17

The MC simulations are performed for the N = 7-17 transition-metal clusters without any confinement by assuming the Gupta potential in (2.5) [2.14]. In what follows, we shall use r_0 and U in (2.5) as units of distance and energy, respectively. Figure 2.23a displays the internal energy per atom, E, and the specific heat C, calculated from energy fluctuations, as given in (2.20). The peak of C corresponds to the change of curvature of E at $T \simeq 0.03$. It is rather broad and shows scattering of the calculated points on the low-temperature side. In Fig.2.23b, the averaged bond length L

$$L = 2 \sum_{ij} \frac{\langle r_{ij} \rangle}{N(N-1)}, \tag{2.21}$$

and the bond-length fluctuations δ given in (2.2) are depicted for the same system. The bond-length L undergoes a gradual change of slope starting from the temperature where δ begins to jump ($T \simeq 0.02$) and ending at the temperature where δ saturates. The latter temperature corresponds to the one giving the peak of C. Figure 2.24a and b illustrate similar plots of E,C,L and δ for the N = 17 cluster. The qualitative behaviors are quite similar to those for the N = 13 cluster except for C which shows no peak in the temperature region studied.

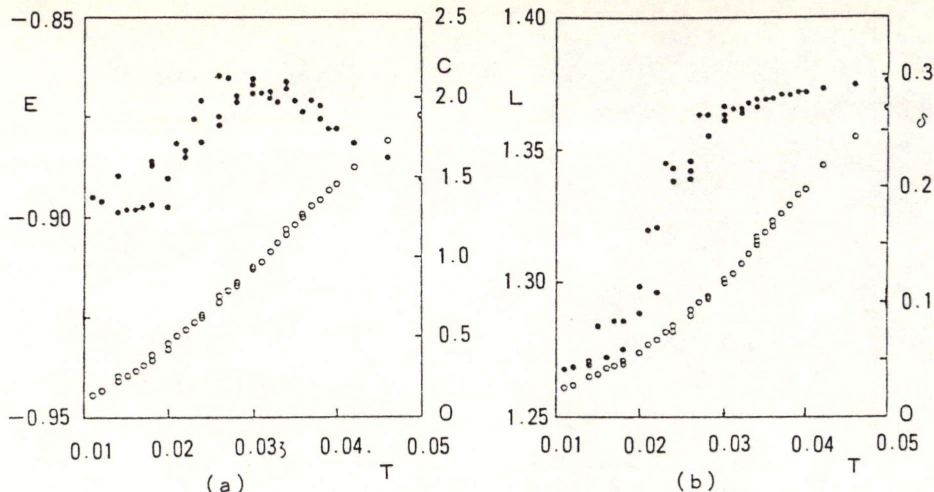

Fig.2.23. **(a)** The internal energy E (∘) and specific heat C (•) vs. T, **(b)** mean bond-length L (∘) and bond-length fluctuations δ (•) vs. T for N = 13 [2.14]

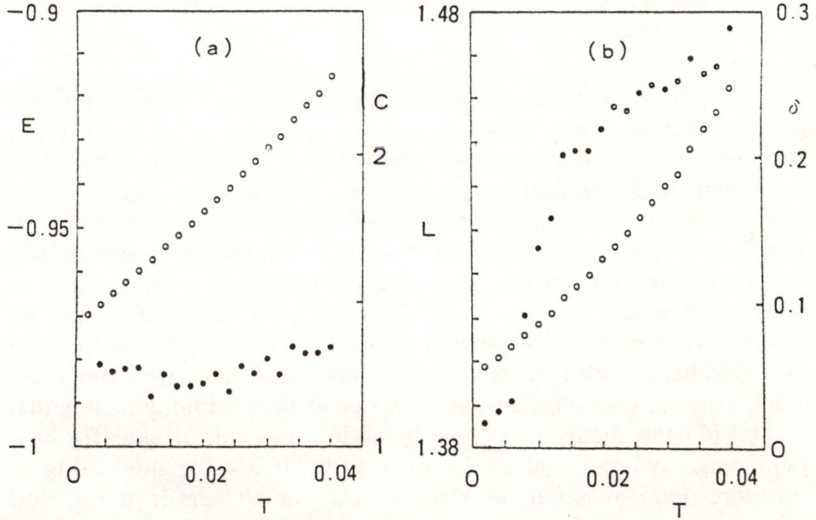

Fig.2.24. **(a)** E (∘) and C (•) vs. T, **(b)** L (∘) and δ (•) vs. T for N = 17 [2.14]

The characteristic temperatures found in the MC simulations for the N = 7-17 clusters, are summarized in Fig.2.25, where the lower- and upper-limiting temperatures of the jump in δ and the temperature at the maximum of C are indicated. In the figure, the results of the Lennard-Jones clusters [2.17, 18], are also displayed for comparison: only the temperatures at which δ rises sharply are indicated. For small sizes, our potential and the Lennard-Jones one give the same behavior, but there is a small deviation for large sizes. This may be due to the many-body effect of our potential enhanced when N is large.

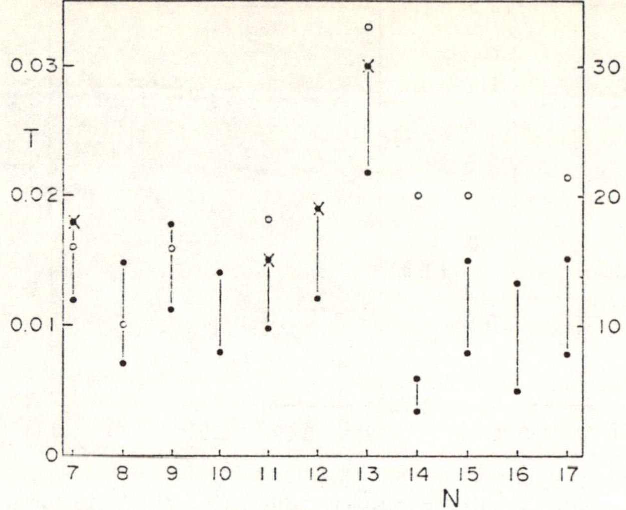

Fig.2.25. Characteristic temperatures versus N. Vertical bars limited by black dots are limits of the jump of δ, and crosses indicate the peak temperatures of C. The results for Lennard-Jones clusters [2.17, 18] (in degrees Kelvin) are shown for comparison by open circles with the right scale [2.14]

The most interesting point of the results seen in Fig.2.25 is the fact that the peak of C is observed only for N = 7, 11, 12 and 13. In connection with this, we remark here that the structure of the N = 7 cluster in the solid phase is a pentagonal bipyramid, the structure of the N = 11 cluster at low temperature consists of two pentagonal rings of five atoms plus one atom on the top (Fig.2.26a), and the structure of N = 12 is that of the icosahedral N = 13 cluster minus one on the top. It is also interesting to remark that the low-temperature structure of N = 8, 9, 10, 14, 15, 16 and 17 showing no peak of C have no pentagonal symmetry (Fig.2.26). It is suggested that the transition-metal clusters studied here may be classified into two categories; (1) the stable clusters showing a peak of specific heat without pentagonal symmetry, and (2) the metastable clusters having no peak of specific heat without pentagonal symmetry at low temperatures. It is quite interesting to examine in more detail whether the structures of the clusters in the second category displayed in Fig.2.26 are their ground states or not.

2.4.3 Effect of Magnetic Interactions

In this subsection, we point out that inclusion of a simple magnetic interaction, in addition to the cohesive interaction as given in (2.5), to simulate magnetic transition-metal clusters may exert strong effects on their specific heat coming from the energy fluctuation due to atomic motion, although the magnitude of the magnetic interaction is quite small as compared with that of the cohesive interaction. Furthermore, we demonstrate that the inclusion of the magnetic interaction induces steep magnetic transitions as found in the magnetic bulk, even in small clusters of N = 7 to 17. The in-

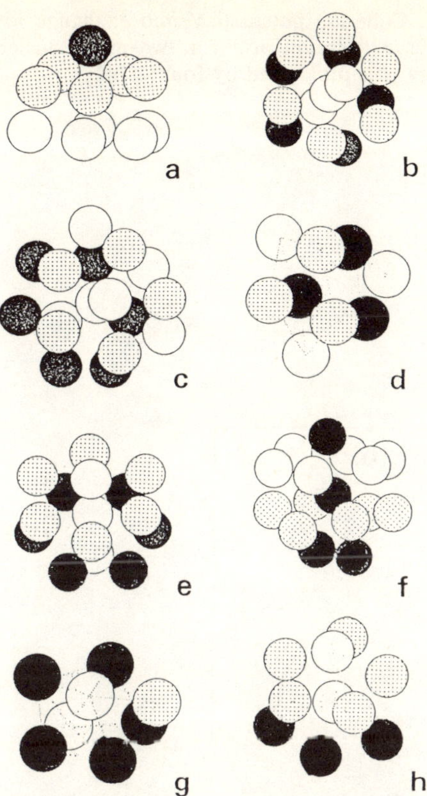

Fig.2.26a-h. Structures at T = 0.002. Particular symmetries are highlighted by sphere colors; (**a**) N = 11, (**b**) N = 14, (**c**) N = 17, (**d**) N = 9, (**e**) N = 15, (**f**) N = 16, (**g**) N = 8 (agray circle and a black circle behind it have equivalent positions with respect to the four black circles), (**h**) N = 10 (the structure of N = 7 plus three black circles) [2.14]

vestigations are made as a continuation of the MC studies described in the previous subsection [2.14]. Arguments on the quantum mechanical origin of magnetism in magnetic metal clusters belong to a different subject and shall be discussed in Sect.4.3. Here we assume microscopic magnetic moments to be localized on the constituent atoms of the clusters and to interact ferromagnetically with each other. As discussed in Sect.4.3, this model is somewhat justified by a recent microscopic calculation of the electronic structure of small transition-metal clusters [2.19].

The magnetic interaction to be assumed is

$$H_{mag} = -\sum_{ij} J(r_{ij}) s_i \cdot s_j \ , \tag{2.22}$$

where the spin s_i at the ith atom is assumed, for simplicity, to be Ising-like with the values ±1, and the $J(r_{ij})$ to be

$$J(r_{ij}) = J_0 \exp(-\alpha r_{ij}) \ . \tag{2.23}$$

In (2.23), J_0 and α are adjustable parameters. In what follows, we choose arbitrarily J_0 and α to be 5 and 6.9, that gives the ratio of the magnetic to

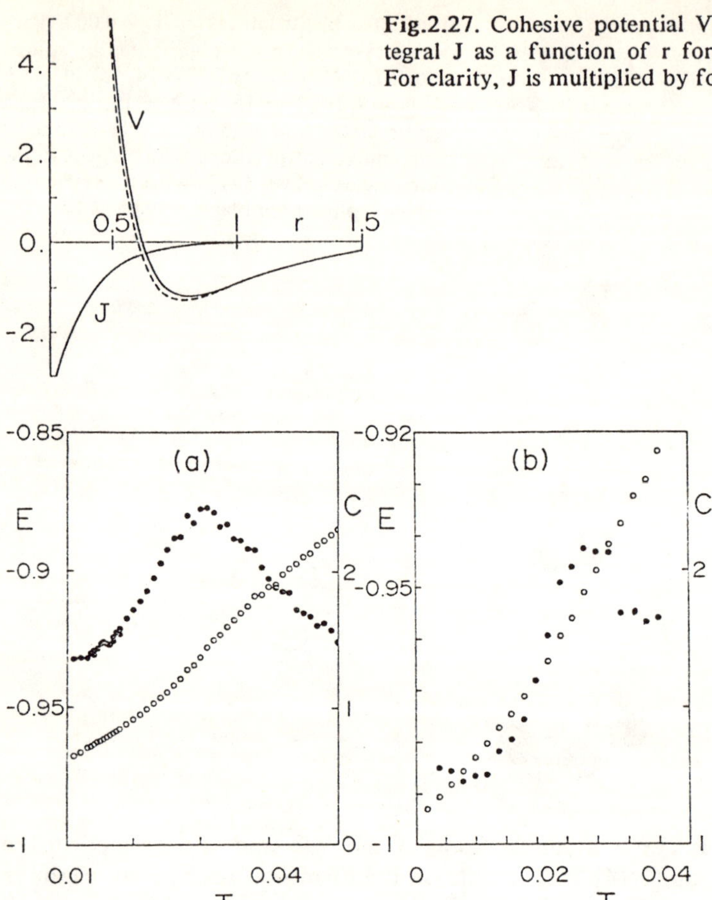

Fig.2.27. Cohesive potential V and exchange integral J as a function of r for two-atom cluster. For clarity, J is multiplied by four [2.14]

Fig.2.28. Internal energy E (○) and specific heat (•) versus T for (**a**) N = 13 and (**b**) N = 17 [2.14]

the cohesive energies at equilibrium distance r_0 to be 0.5%. Note that this ratio in the bulk of transition-metals is estimated to be around 0.1% from the Curie temperatures and the melting temperatures by using the mean-field approximation. The magnetic interaction J(r) is compared with the cohesive potential of (2.5) in Fig.2.27.

MC simulations similar to those without the magnetic interaction are performed by taking into account additional freedom of motion due to spins. The calculated internal energy E and specific heat C as a function of temperatures for the N = 13 and 17 clusters are depicted in Fig.2.28. Comparing Figs.2.28a with 2.23a, one observes that inclusion of the magnetic interaction only suppresses scattering of points and makes the peak of C sharper for N = 13 belonging to the stable category (N = 7,11,12,13). On the other hand, comparing Fig.2.28b with Fig.2.24a, one notes a surprising effect of the magnetic interaction in the appearance of the peak of specific heat which does not exist in the non-magnetic case for N = 17 belonging to the metastable category (N = 8,9,14,15,16,17). The reason for this effect is not clear as yet, but we feel that this would be due to the suppression of the structural fluctuation, which is large for clusters in the metastable category, by the magnetic interaction.

One of the most interesting results in the magnetic case is the observation of steep magnetic transitions in small clusters. Magnetization M is calculated from

$$\langle M \rangle = \frac{1}{\Delta} \int_{t_0}^{t_0+\Delta} \left[\frac{1}{N} \sum_i m_i(t) \right] dt \tag{2.24}$$

where t_0 is the time when thermal equilibrium is established, $m_i(t)$ the magnetic moment on the ith atom at time t, Δ the appropriate time interval to make $\langle M \rangle$ to be convergent. The magnetic susceptibility χ is calculated from the fluctuation of magnetization as

$$\chi = (\langle M^2 \rangle - \langle M \rangle^2)/(NT) \ . \tag{2.25}$$

Figure 2.29 exhibits the calculated M and χ as a function of T for the N = 7 and 17 clusters. The jump of χ at the magnetic transition from the low temperature looks discontinuous. On the high-temperature side it shows a smooth behavior. We note that no peak in C is observed at the magnetic transition. This would be due to smallness of the magnetic energy as compared with the cohesive energy. The steepness of the magnetic transition for clusters as small as N = 7 seems to be a consequence of the interplay between the magnetic and cohesive interactions. However, the reason has not yet been clarified.

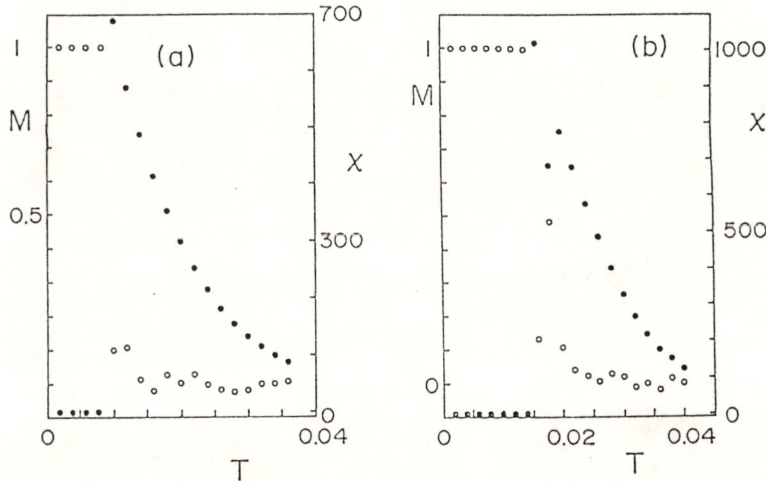

Fig.2.29a,b. Magnetization M (o) and magnetic susceptibility χ (•) versus T for (a) N = 7 and (b) N = 17 [2.14]

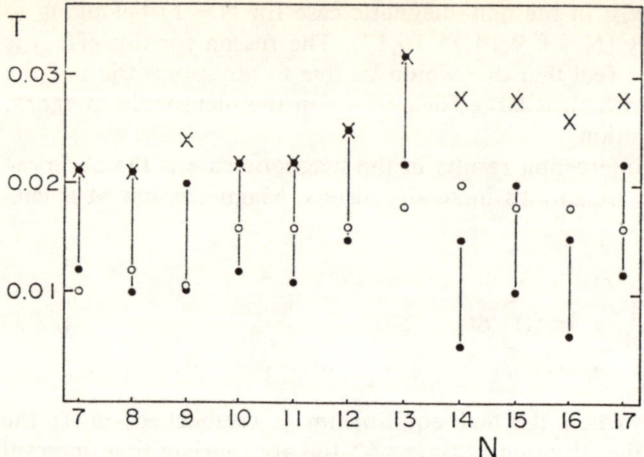

Fig.2.30. Characteristic temperatures versus N in the presence of the magnetic interaction: Vertical bars limited by full circles show the range of the jump of δ, crosses the peak of C, and open circles the magnetic transitions [2.14]

We illustrate in Fig.2.30 the magnetic transition temperature and the melting temperature as a function of N. The peak of C is observed for all the sizes studied. The magnetic transition temperature is always below the temperature giving the peak of C.

3. Shell Structure of Metal Clusters

As pointed out in Sect. 1.2.2, it is not an exaggeration to mention that finding of the shell structure in alkali- and noble-metal microclusters has opened a new field of microcluster physics. In this chapter we first elucidate theoretically the shell structure of metal clusters by using a simple model. The shell structure is expected to induce a deformation of the clusters with open shells and play an important role in the fragmentation of microclusters as found in the nuclear fission. These shell effects are the main subjects of this chapter.

3.1 Magic Numbers

As mentioned in Sect. 1.2.2, anomalies of the abundance have been observed at specific sizes N called *magic numbers* in the mass spectra of a microcluster beam of alkali- and noble-metal elements, indicating the microclusters of these sizes to be relatively stable as compared with those of the neighboring sizes. We have already displayed in Fig. 1.5 the observed mass spectrum of a Na cluster beam with the magic numbers N = 8, 20, 40, 58 and 92 [3.1]. In this section, we depict in Fig. 3.1 the mass spectra of $Ag_N^{+,-}$ cluster ions directly produced by bombardment of a silver target with Xe ions of 10 keV energy [3.2]. In the figure, the origin of the abscissa for Ag_N^- is shifted towards the right by two cluster sizes from that for Ag_N^+, in order to emphasize a similarity between the mass spectra of Ag_N^+ and Ag_N^-. In both the spectra, we see two types of anomalies. One is the odd-even alternation observed up to N = 30÷40, and another, the discontinuous variation at the magic numbers. The magic numbers of the negative cluster are N = 1, 7, 17, 19, 33, 39 and 57, and those of the positive cluster N = 3, 9, 19, 21, 35, 41 and 59. Such an observation tells us that the magic number is determined by the specific number N_e of valence electrons but not by that of atoms N in microclusters.

The magic numbers are observed in cluster ions of the noble-metal elements Cu, Ag and Au. They are listed in Table 3.1. As already mentioned in Sect. 1.2.2, the magic numbers are the numbers of valence electrons at the closed-shell configuration, where the shells arise from the individual motion of an electron in a spherically symmetric effective potential. The shell structure in three types of simple effective potential without the spin-orbit interaction was already depicted in Fig. 1.6 [3.3], and explanations of

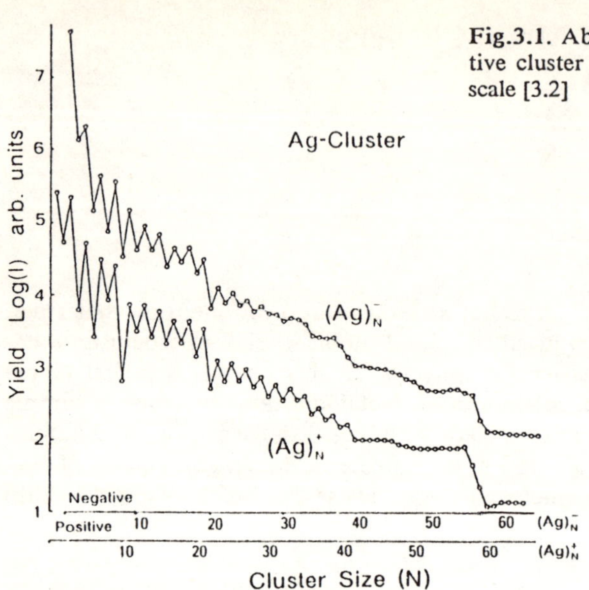

Fig.3.1. Abundance of positive and negative cluster ions of silver on a logarithmic scale [3.2]

Table 3.1. Magic number of noble-metal cluster ions (○: clearly observed, Δ: vaguely observed, ×: absent) [3.2]

(N_1-1) or (N_2+1)	Closed shells	$Cu^+_{N_1}$	$Cu^-_{N_2}$	$Ag^+_{N_1}$	$Ag^-_{N_2}$	$Au^+_{N_1}$	$Au^-_{N_2}$
2	1s	○		○	○	○	×
8	1p	○	○	○	○	○	×
18	1d	Δ	Δ	Δ	Δ	Δ	○
20	2s	○	○	○	○	○	×
34	1f	○	○	○	○	○	○
40	2p	○	○	○	○	×	×
58	1g	○	○	○	○	○	○
92	3s	○	○	○	○	○	○
138	3p	○	Δ	○	○	○	Δ
198	4s			○	Δ		

the figure are given in Sect.1.2.2. The closed shells indicated in Table 3.1 may be considered to be those in the square well potential with an infinite wall, as shown in Fig.1.6. The reason why the use of such a simple potential can explain the order of the shells determined from the observed magic numbers is discussed in Sect.3.2.

Generally speaking, an inhomogeneous distribution of energy levels of a single Fermi particle, as schematically illustrated in Fig.3.2, may be re-

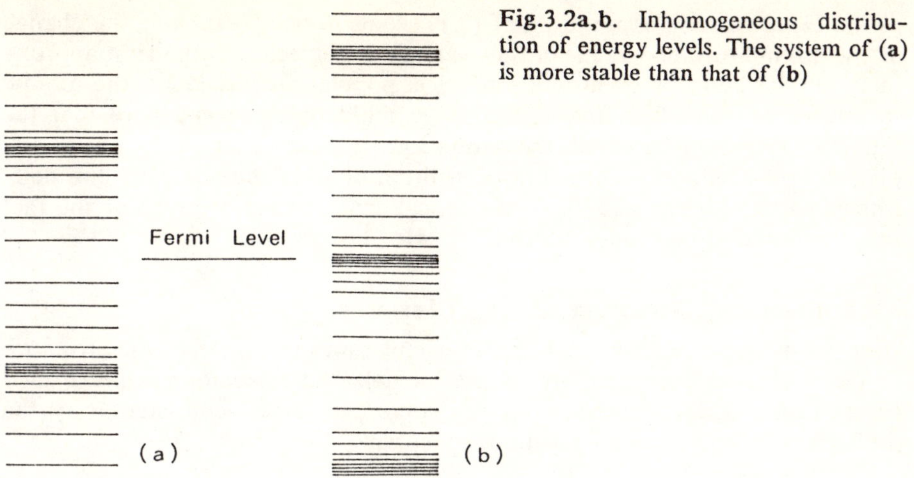

Fig.3.2a,b. Inhomogeneous distribution of energy levels. The system of (**a**) is more stable than that of (**b**)

garded as a shell structure. The magic number of this system is given by the specific number of particles giving minimum degree of degeneracy, taking into account near degeneracy in an appropriate way, when the particles are accommodated in the single-particle energy levels. In the case of Fig.3.2a more energy levels are found in lower energies as compared with the case of Fig.3.2b, so that one can easily judge that (a) is more stable than (b): Note that the number of energy levels below the Fermi level is the same in both cases.

The magic numbers arising from the origins other than the shell structure, such as geometry of atomic structure and electron correlation, will be discussed in later chapters.

3.2 The Jellium Model

For discussing the electronic structure of alkali-metal crystals, where the nearly-free-electron picture is valid, the jellium model replacing core ions by a continuous background of positive charges is known to be a good approximation [3.4]. Similarly, the spherical jellium model is successfully applied to elucidating the electronic structure of alkali metal clusters [3.5-7]. In the jellium model we assume that the density of core ions is constant, being independent of the number of constituent atoms of clusters. Then the radius R_0 of a jellium sphere is given as

$$R_0 = r_s N^{1/3} , \tag{3.1}$$

where r_s is the Wigner-Seitz radius of the corresponding crystal. Constancy of nucleon density, called the *saturation property of nuclei*, is well known to provide the basis of the liquid-drop model giving Bethe-Weizsächer mass fomula. For metal clusters, no theoretical justification of (3.1) has yet been given. However, the assumption of (3.1) seems to be agreeable if the change of the atomic distance at the cluster surface is neglected to the first approximation: the concept of atomic radius is popular in discussing the atomic structures of molecules and crystals. It should be remarked here that the liquid-drop model based on the saturation property is applied [3.8] to explaining the frequency spectrum of gallium clusters measured by the neutron inelastic scattering [3.9]. A theoretical confirmation is given of the fact that the liquid-drop model is valid for ^{4}He clusters [3.10].

3.2.1 Electronic Structure of Na_N Clusters

Here an example is discussed of the calculation of the electronic structure of metal clusters based on the jellium model. The calculation is performed for neutral sodium clusters, Na_N, by solving in a self-consistent way the problem of N-electrons in a jellium potential:

$$v_+(\mathbf{r}) = \begin{cases} -(N/2R_0)[3-(r/R_0)^2]\,, & \text{for } r < R_0 \\ -N/r & \text{for } r \geq R_0\,. \end{cases} \tag{3.2}$$

To solve the problem of N electrons we use the spin-polarized Local-Density-Function (LDF) approximation [3.11]; namely, we solve self-consistently the Kohn-Sham equation [3.12]

$$[-\tfrac{1}{2}\nabla^2 + v_{eff}(\mathbf{r};\sigma)]\psi_{i\sigma}(\mathbf{r}) = \epsilon_{i\sigma}\psi_{i\sigma}(\mathbf{r}) \tag{3.3}$$

where effective potential $v_{eff}(\mathbf{r};\sigma)$ is given as

$$v_{eff}(\mathbf{r};\sigma) = v_+(\mathbf{r}) + \int d^3\mathbf{r}'\rho_-(\mathbf{r}')/|\mathbf{r}-\mathbf{r}'| + \mu_{xc}^{\sigma}(\rho\uparrow(\mathbf{r}),\rho\downarrow(\mathbf{r}))\,. \tag{3.4}$$

In (3.4), the last term is given in terms of the exchange-correlation energy per electron $\epsilon_{xc}(\rho\uparrow,\rho\downarrow)$ as

$$\mu_{xc}^{\sigma}(\rho\uparrow,\rho\downarrow) = \frac{\partial}{\partial\rho_\sigma}[\rho_-\epsilon_{xc}(\rho\uparrow,\rho\downarrow)]\,, \tag{3.5}$$

and $\rho_\sigma(\sigma = \uparrow,\downarrow)$ and ρ_- are defined by

$$\rho_\sigma(\mathbf{r}) = \sum_i \rho_{i\sigma}(\mathbf{r}) = \sum_i f_{i\sigma}|\psi_{i\sigma}(\mathbf{r})|^2\,, \tag{3.6}$$

$$\rho_-(\mathbf{r}) = \sum_\sigma \rho_\sigma(\mathbf{r})\,,$$

where $\Sigma_i f_{i\sigma} = N_\sigma$, $N_\uparrow + N_\downarrow = N$, and $0 \leq f_{i\sigma} \leq 1$: N_σ is the total number of electrons of spin σ. By using the calculated spin-polarized electron densities

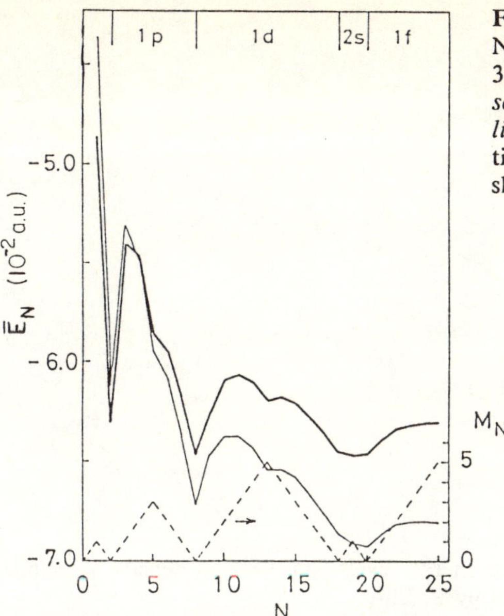

Fig.3.3. Total energy per atom, $\overline{E}_N$, for Na_N clusters with jellium spheres of r_s = 3.93 a.u. calculated by use of LDF (*thin solid line*) and LDF-SIC (*thick solid line*) approximations. The spin polarization M_N of the ground state is also shown (*thin dashed line*) [3.7]

$\rho_\sigma(\mathbf{r})$, the total energy of the cluster per atom, $\overline{E}_N$, for N = 1÷25 is obtained, as depicted in Fig.3.3. It is well known that the LDF approximation may be improved by taking into account the Self-Interaction Correction (SIC) [3.13]. The result obtained by using the SIC version is also exhibited in Fig.3.3, together with the spin polarization, $M_N = N_\uparrow - N_\downarrow$. The line of the calculated total energy shows downward cusps at N = 2, 5, 8, 13, 18, 19 and 20, among which N = 2, 8, 18 and 20 are, respectively, the shell-closing numbers of the 1s, 1p, 1d and 2s shells while N = 5, 13 and 19 are the numbers giving the half-filled 1p, 1d and 2s shells, respectively, with high-spin polarization, as indicated by dashed line.

One of the interesting features of Fig.3.3 is that the order of the energy levels of our spherical jellium model is similar to that of the spherical square well (Fig.1.6). This is due to a similarity of the effective potential in (3.4) to the square well potential, as depicted in Fig.3.4. Detailed examination shows that the effective potential arises solely from the exchange-correlation potential, which is almost homogeneous inside the jellium sphere where the electron is nearly free and quickly goes to zero outside where the electron is absent.

The calculated total energy per atom in Fig.3.3 seems to consist of two parts. One is monotonously decreasing and approaching a fixed value as N increases. Another is oscillating with a decreasing amplitude as N is increasing. We shall show in Sect.3.3 that the former corresponds to the energy of a liquid droplet and the latter comes from the shell effect.

In order to examine the quantitative validity of the LDF calculation based on the jellium model, the ionization potential I_N of the same system is computed (Fig.3.5). The calculated I_N displays a remarkable oscillation

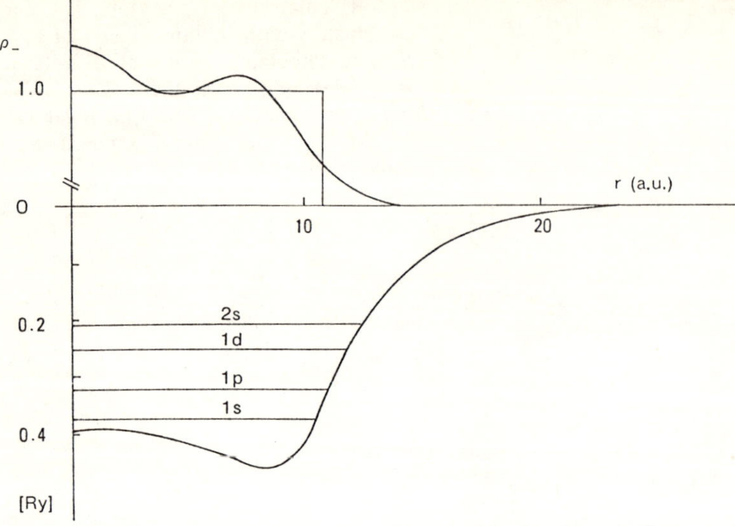

Fig.3.4. Effective potential, one-electron energy levels and electron density ρ_- calculated for Na_{20} by using the jellium model by Y. Ishii

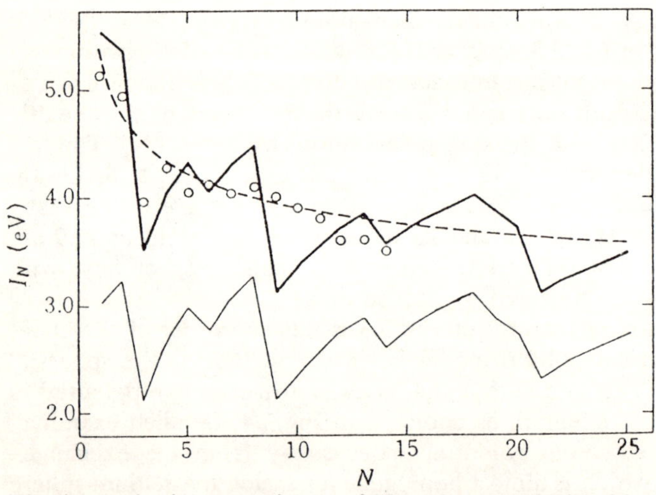

Fig.3.5. Ionization potential I_N of Na_N clusters calculated by using the jellium model with LDF (*thin line*) and the LDF-SIC (*thick line*) approximation. (Void circles: experimental [3.15]. Dashed curve: classical result [3.16]). [3.7]

corresponding to the shell structure, while the experimental result does not. However, the recent experiment by W.A. Saunders for K_N clusters clearly reveal an oscillatory shell effect, as shown in Fig.3.6 [3.4].

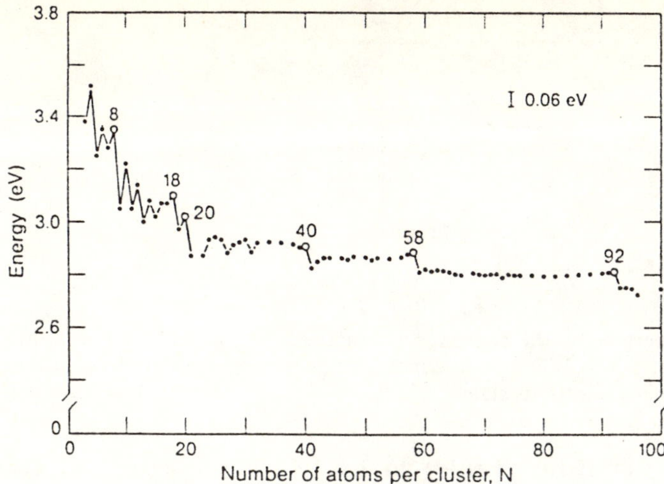

Fig.3.6. Experimental values of ionization potential I_N of K_N clusters measured by *Saunders* [3.14]

3.2.2 Spin Polarization and Multiplet Formation

In this subsection we consider a metal cluster with an unfilled electron shell. We assume that the interaction energy between the electrons in the shell is independent of magnetic quantum number m. Denoting the interaction energy for parallel and antiparallel spin pairs as $E_{\uparrow\uparrow}$ and $E_{\uparrow\downarrow}$, respectively, the total energy of the electron-electron interaction in the shell, E_{int}, is given by

$$E_{int} = E_{\uparrow\uparrow} \sum_{\sigma} n_\sigma(n_\sigma - 1)/2 + E_{\uparrow\downarrow} n_\uparrow n_\downarrow$$

$$= E_{\uparrow\uparrow} n(n-1)/2 + \Delta\epsilon_{exch} n_\uparrow n_\downarrow \ , \tag{3.7}$$

where n_σ is the number of the electrons of spin σ in the shell, and $\Delta\epsilon_{exch} = E_{\uparrow\downarrow} - E_{\uparrow\uparrow}$ which is positive in general. Therefore, E_{int} is lowest when the spin polarization attains its maximum value. Within either the LDF or the LDF-SIC approximation, $\Delta\epsilon_{exch}$ for the 1d electrons in sodium clusters is estimated as

$$\Delta\epsilon_{exch} = (3.5 \pm 0.5) \cdot 10^{-3} \text{ a.u.} \ . \tag{3.8}$$

In addition to the stabilization by the spin polarization, the ground state with the maximum spin polarization is further stabilized by the multiplets formation in the cases of d^2, d^3, d^7 and d^8 configurations (N = 10,11,15 and 16). This situation is schematically illustrated in Fig.3.7. The stabilization energy due to the multiplets formation in all these cases is obtained in terms of Slater integrals as

$$\Delta\epsilon_{mult} = [9F^2(dd) - 5F^4(dd)]/98 \ . \tag{3.9}$$

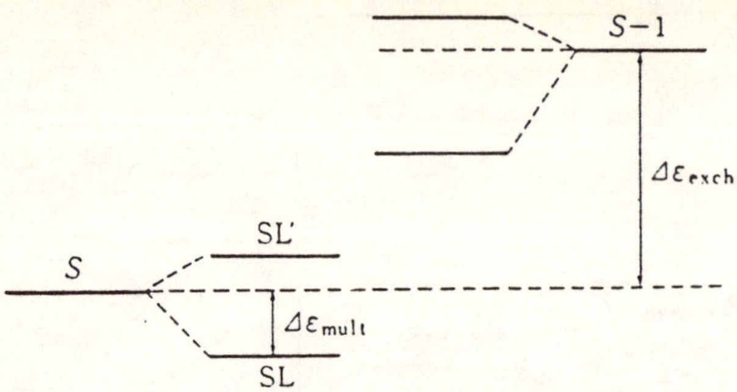

Fig.3.7. Stabilization energies by the exchange interaction, $\Delta\epsilon_{exch}$, and by the multiplets formation, $\Delta\epsilon_{mult}$. Symbols S and L denote, respectively, the spin and orbital angular momenta of a many-electrons state

If we assume the ratio $F^2(dd)/F^4(dd)$ to be 1.4 [3.7] as found by the straightforward integration by using the 1d wave function of the jellium sphere, multiplet stabilization energy $\Delta\epsilon_{mult}$ is estimated as

$$\Delta\epsilon_{mult} = 0.45\Delta_{exch} = (1.6\pm0.2)\cdot10^{-3}\ \text{a.u.}\ , \tag{3.10}$$

which shows that the multiplets formation gives minor but non-negligible effects as compared with the spin polarization.

3.2.3 Nonspherical Perturbation

It is well known that the spin polarization of a many-electrons system of spherical symmetry is quenched by nonspherical perturbation such as a cubic field, low-symmetry fields etc. In this subsection we study effects of nonspherical potential which has to exist in real clusters.

The nonspherical potential $v_{ion}(\mathbf{r})$ is constructed by superposing an appropriate model potential $v(\mathbf{r}-\mathbf{R}_i)$ as

$$v_{ion}(\mathbf{r}) = \sum_i v(\mathbf{r}-\mathbf{R}_i)\ , \tag{3.11}$$

where $\mathbf{R}_i$ is the position vector of the ith atom in a cluster. The following two cases of the model potential are studied for the Na_{13} cluster; the first case is that where the model potential is the jellium potential for N = 1,

$$v_{JEL}(\mathbf{r}) = \begin{cases} -[3-(r/r_s)^2]/(2r_s) & \text{for } r \le r_s \\ -1/r & \text{for } r > r_s\ . \end{cases} \tag{3.12}$$

The second is the case where it is the model pseudopotential [3.17, 18],

$$v_{AL}(\mathbf{r}) = \begin{cases} 0 & \text{for } r \le r_c \\ -1/r & \text{for } r > r_c\ , \end{cases} \tag{3.13}$$

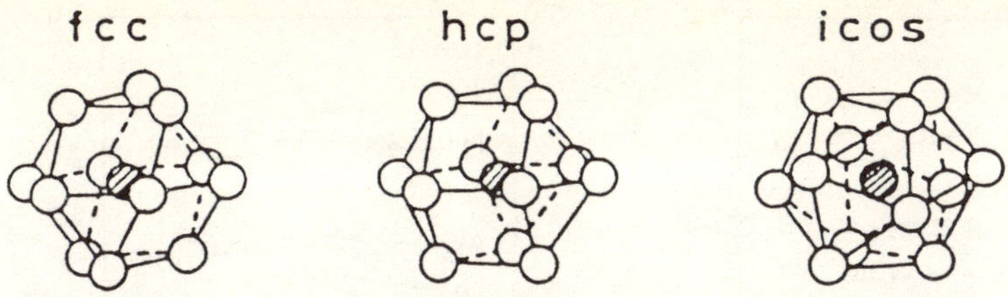

Fig.3.8. Three spatial configurations of atoms with high symmetry in the N = 13 cluster

where r_c = 1.67 a.u. The nonspherical perturbation is given by

$$\Delta v(\mathbf{r}) = v_{ion}(\mathbf{r}) - v_{+}(\mathbf{r}) , \tag{3.14}$$

$v_{+}(\mathbf{r})$ being the potential of a spherical jellium.

For the N=13 cluster, the following spatial configurations of the constituent metal ions with high symmetry are conceivable; face-centered-cubic (fcc), hexagonal-closed-packed (hcp) and icosahedral (icos) configurations (Fig.3.8). The calculations of level splitting in these configurations are performed by the use of two kinds of the model potentials in (3.12 and 13). The results are shown in Fig.3.9. In any configuration, the splitting is smaller than the level separation of the spherical jellium except for that between the 1d and 2s levels. Such a result is due to a good screening of the core ion potentials by valence electrons resulting in small nonspherical perturbation in sodium clusters. It is interesting to study the level splitting in noble-metal clusters, where the screening of the core potential does not seem to be good, as in alkali-metal clusters.

The competition between the exchange energy and the nonspherical perturbation energy determines whether or not the Hund rule is valid for the N=13 alkali-metal cluster with the half-filled 1d shell. The energy difference between the low-spin (S = 1/2) and the high-spin (S = 5/2) states in the fcc configuration is

$$E(S=1/2) - E(S=5/2) = 6\Delta\epsilon_{exch} - 2\Delta_{fcc} , \tag{3.15}$$

where Δ_{fcc} is the d-level splitting in the fcc configuration. Using (3.8) and the calculated value of $\Delta_{fcc} = 2.0 \cdot 10^{-2}$ a.u., we find the energy difference in (3.15) is negative. This shows that the Hund rule is broken by the nonspherical perturbation in the Na_{13} cluster of the fcc configuration.

3.3 Theory of Shell Correction

In the previous section we pointed out the importance of the saturation property, on which the liquid-drop model is based, and the shell effects in metal microclusters. The liquid-drop and the shell models shed light on two

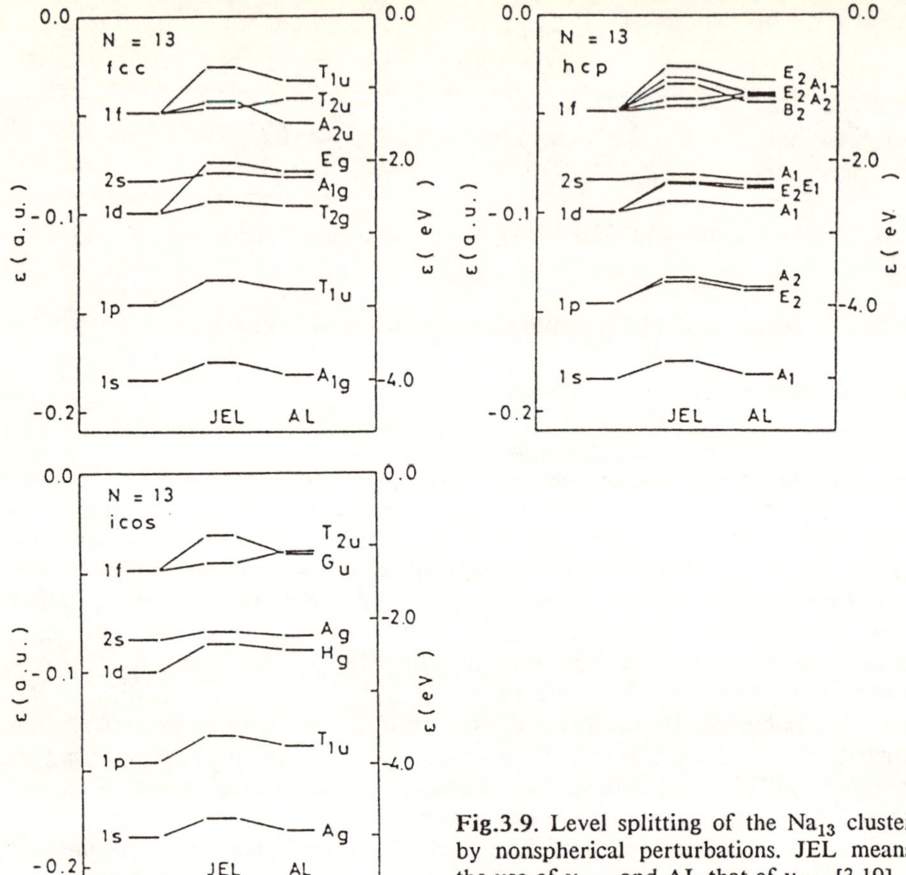

Fig.3.9. Level splitting of the Na_{13} cluster by nonspherical perturbations. JEL means the use of v_{JEL} and AL that of v_{AL}. [3.19]

different aspects of a small many-body system, collective motion of the constituent particles and one-particle motion in an effective potential. A unified treatment of these two models was given in the field of nuclear physics [3.20]. The treatment, however, is not adapted to studying large deformations such as those important for nuclear fission etc.

The method of treating large deformations as well as the shell effect of heavy nuclei has been developed [3.21]. In what follows, we introduce this theory for the purpose of applying it to the studies of largely deformed metastable states (fission isomers) and fission of metal clusters. In the theory, the main part of the total energy of a deformed nucleus is treated by using the liquid-drop model and the shell effect is taken into account as a small correction to it. This idea is based on the fact that the total energy is approximately proportional to N while the shell energy correction to $N^{\alpha}(\alpha<1)$: $\alpha = 1/3$ for the harmonic oscillator model, as will be demonstrated in Sect.3.3.2 and $\alpha = -1/3$ for the spherical well.

3.3.1 Essence of the Theory

To explain the essence of the theory of shell corrections, we adopt the Hartree-Fock scheme giving the total energy of a N-electrons system, E^{HF}, as

$$E^{HF} = \sum_{\nu \in F} \langle \nu | t | \nu \rangle + \frac{1}{2} \sum_{\nu,\mu \in F} \langle \nu\mu | v_{12}(1-p) | \nu\mu \rangle , \tag{3.16}$$

where $\nu \in F$ means the occupied spin-orbital ν below the Fermi level ϵ_F, t a one-electron operator representing kinetic plus potential energies, v_{12} the Coulomb interaction operator, and p the permutation operator. Within this theoretical scheme, the orbital energy of ν is given by

$$\epsilon_\nu = \langle \nu | t | \nu \rangle + \sum_{\mu \in F} \langle \nu\mu | v_{12}(1-p) | \nu\mu \rangle . \tag{3.17}$$

It is evident that

$$E^{HF} \neq \sum_{\nu \in F} \epsilon_\nu (\equiv U) , \tag{3.18}$$

as the Coulomb interaction is counted twice in U. For replacing the sums in (3.16 and 17) by integrals as

$$\sum_{\nu \in F} \ldots \rightarrow \int_{-\infty}^{\epsilon_F} d\epsilon g(\epsilon) \ldots , \tag{3.19}$$

we introduce level-density function $g(\epsilon)$ defined as

$$g(\epsilon) = \sum_{\nu} \delta(\epsilon - \epsilon_\nu) , \tag{3.20}$$

where $\delta(\epsilon-\epsilon_\nu)$ is the delta function. Further, we introduce the smoothed level-density function $\tilde{g}(\epsilon)$, which is obtained by broadening the energy levels ϵ_ν appropriately. The method for obtaining $\tilde{g}(\epsilon)$ will become explicit for an example in the next subsection. A central point of the theory of shell correction is to assume the integral

$$\int_{-\infty}^{\epsilon_F} d\epsilon \Delta g(\epsilon) \ldots \tag{3.21}$$

with $\Delta g(\epsilon) \equiv g(\epsilon)-\tilde{g}(\epsilon)$, to be a small quantity of the first order, as compared with the corresponding integral with $\Delta g(\epsilon)$ being replaced by $g(\epsilon)$.

It is straightforward to obtain the following expression for the total energy E_{tot} of the system, to the approximation of neglecting small quantities of higher order:

$$E_{tot} = \tilde{E} + \Delta U \,, \tag{3.22}$$

$$\tilde{E} - E_{core} \equiv \tilde{E}^{HF} = \int_{-\infty}^{\epsilon_F} d\epsilon \tilde{g}(\epsilon) \langle \epsilon | t | \epsilon \rangle + \frac{1}{2} \int_{-\infty}^{\epsilon_F} d\epsilon \int_{-\infty}^{\epsilon_F} d\epsilon' \tilde{g}(\epsilon) \tilde{g}(\epsilon')$$

$$\times \langle \epsilon\epsilon' | v_{12}(1-p) | \epsilon\epsilon' \rangle \,, \tag{3.23}$$

$$\Delta U = \int_{-\infty}^{\epsilon_F} d\epsilon \Delta g(\epsilon) \epsilon = U - \tilde{U} \,, \tag{3.24}$$

$$\tilde{U} = \int_{-\infty}^{\epsilon_F} d\epsilon \tilde{g}(\epsilon) \epsilon \,. \tag{3.25}$$

In (3.23), E_{core} is the total energy of ion cores in the absence of valence electrons. Since $\tilde{E}^{HF}$ is the total energy of a N-electrons system with the smoothed level-density function $\tilde{g}(\epsilon)$, $\tilde{E}$ may be regarded as the total energy of the cluster without the shell effect. For heavy nuclei where Bosons corresponding to ion cores are absent, such a total energy is assumed to be that of a classical or semiclassical liquid-droplet. In what follows, we also use a similar liquid-drop model for metal microclusters, assuming that ion cores may be regarded as in incompressible jellium.

The expression in (3.22) tells us that the total energy of a metal microcluster of large N is given by the sum of a main classical energy, $\tilde{E}$, and a small quantum correction, ΔU. The second part called shell corrections is a small quantity of the first order, as compared with the first term as assumed at the beginning. In the next subsection, we shall show that this is really the case for large N when an effective potential for a valence electron is of the harmonic oscillator type.

It should be emphasized that the theory of shell corrections as expressed in (3.22), requires no complicated calculation of the electron-electron interaction, thus making possible a simple but quantitative estimation of E_{tot} as long as an appropriate classical model such as the liquid-drop model is used for the calculation of $\tilde{E}$ and an appropriate effective potential for the calculation of the shell correction. The first-order small quantity of the shell correction is not so sensitive to the assumed form of the effective potential.

3.3.2 Shell Correction for the Harmonic-Oscillator Model

In order to understand the theory of shell correction more clearly, we calculate here the shell correction as an example by assuming the effective potential for the motion of an individual valence electron to be of the harmonic-oscillator type:

$$v(r) = \frac{1}{2}\omega_0{}^2(x^2 + y^2 + z^2) . \tag{3.26}$$

The energy eigenvalues for the electron in this potential is known to be

$$\epsilon_n = \omega_0(n + 3/2) \quad \text{(n denoting integers)} \tag{3.27}$$

with the degeneracy

$$g_n = (n + 1)(n + 2) . \tag{3.28}$$

Now, broadening the energy levels given by (3.27), we define a smoothed level-density function $\tilde{g}(\epsilon)$ in such a way that the number of the broadened energy levels in the energy range between ϵ and $\epsilon+d\epsilon$ is expressed by

$$\tilde{g}(\epsilon)d\epsilon = g(n)dn \tag{3.29}$$

where ϵ and $g(n)$ are, respectively, ϵ_n in (3.27) and g_n in (3.28) extended for non-integral number n. Then, by using (3.27, 28), $\tilde{g}(\epsilon)$ is obtained as

$$\tilde{g}(\epsilon) = g(n)\frac{dn}{d\epsilon} = \frac{1}{\omega_0}g(n) = \frac{1}{\omega_0}\left(\frac{\epsilon^2}{\omega_0{}^2} - \frac{1}{4}\right) . \tag{3.30}$$

From the relation

$$N = \int_0^{\epsilon_F} d\epsilon\tilde{g}(\epsilon) , \tag{3.31}$$

the Fermi energy ϵ_F for large N is obtained as

$$\epsilon_F/\omega_0 = (3N)^{1/3} + \frac{1}{4}(3N)^{-1/3} , \tag{3.32}$$

where small quantities of higher order are neglected. Since the cluster radius R_0 is related to the Fermi energy ϵ_F, as shown in Fig.3.10, and to the Wigner-Seitz radius r_s as in (3.1), we obtain for ω_0

$$\omega_0 = 2{\cdot}3^{2/3}r_s{}^{-2}\left[(3N)^{-1/3} + \frac{1}{4}(3N)^{-1}\right] , \tag{3.33}$$

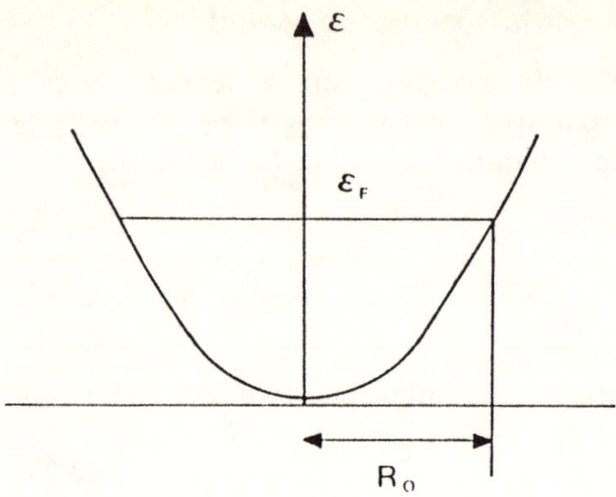

Fig.3.10. Relation between ϵ_F and cluster radius R_0

from (3.32) and

$$R_0 = (2\epsilon_F/\omega_0^2)^{1/2} = N^{1/3} r_s \tag{3.34}$$

expressing ϵ_F in terms of r_s. Inserting (3.33) into (3.32), one obtains

$$\epsilon_F = 2\cdot 3^{2/3} r_s^{-2}\left[1 + \frac{1}{2}(3N)^{-2/3}\right], \tag{3.35}$$

which indicates that ϵ_F approaches a fixed value from the above when N increases. Such a behavior of ϵ_F is independent of the form of the effective potential assumed.

The shell correction ΔU in our problem can be calculated in the following way; from (3.25) combined with (3.30), it is straightforward to obtain

$$\tilde{U} = \frac{\omega_0}{4}\left[(3N)^{4/3} + \frac{1}{2}(3N)^{2/3}\right]. \tag{3.36}$$

To simplify the argument, we calculate U at specific $\overline{N}$ where the degenerate $\bar{n}$th level is completely filled: the amplitude of shell correction $|\Delta U|$ becomes the maximum at $\overline{N}$. By using the relation,

$$\overline{N} = \sum_{n=0}^{\bar{n}} g_n = \frac{1}{3}(\bar{n}+1)(\bar{n}+2)(\bar{n}+3), \tag{3.37}$$

one obtains for large $\overline{N}$

$$\bar{n} = (3\overline{N})^{1/3} - 2 + \frac{1}{3}(3\overline{N})^{-1/3}, \tag{3.38}$$

where small quantities of higher order are neglected. Then, the sum of the orbital energies, U, defined in (3.18) for large $\overline{N}$ is calculated to be

$$U(N=\overline{N}) = \sum_{n=0}^{\bar{n}} \omega_0(n+1)(n+2)(n+3/2)$$

$$= \frac{\omega_0}{4}\left[(3N)^{4/3} + \frac{1}{3}(3\overline{N})^{2/3}\right], \tag{3.39}$$

by using (3.38). Finally, from (3.36, 39), the shell correction ΔU for large $\overline{N}$ is obtained as

$$\Delta U(N=\overline{N}) = -\frac{\omega_0}{24}(3\overline{N})^{2/3} = -\overline{N}^{1/3}/(4r_s^2), \tag{3.40}$$

where small quantities of higher order are neglected.

The important result of our specific model of the harmonic-oscillator potential is that the amplitude of the oscillating ΔU in (3.22) is proportional to $N^{1/3}$ while the main part of $\tilde{E}$ in (3.22) to N, as may be shown by using the liquid-drop model. This result justifies our starting assumption that the integral in (3.21) is a small quantity of first order as compared with that in (3.19). In our model, the amplitude of ΔU is of the order of 1% of $\tilde{E}$ at N = 10^3, which gives the upper bound of the size of microclusters where the quantum-mechanical shell effect is important. It should, however, be noted here that our model may overestimate the shell correction as it gives too much degeneracy to the one-electron energy levels.

3.4 Deformation

As mentioned in Sect.3.1, the observation of the magic numbers corresponding to the shell-closing numbers of valence electrons in a spherical system clearly reveals that at least the microclusters of these magic numbers are spherically symmetric. *Katakuse*'s experiments on both positively and negatively ionized noble-metal clusters [3.2] seems to tell us that the spherical symmetry of the clusters is determined by the number of valence electrons but not by that of ion cores. There still remains a question, however, whether the microclusters of non-magic number are really deformed or not.

3.4.1 Application of the Shell Correction Theory

In order to discuss deformation of microclusters, we apply the theory of shell correction, in which the liquid-drop model is used for calculating the classical energy and the anisotropic harmonic oscillator model for the shell correction. For simplicity, we assume that the deformation is spheroidal as given by

$$(x^2+y^2)e^{\gamma} + z^2 e^{-2\gamma} = R_0^2 \, . \tag{3.41}$$

For this spheroid, the ratio of the length of the rotational axis, the z axis, to that of the perpendicular axis, the x or y axis, is $\exp(3\gamma/2)$ and the volume is independent of deformation parameter γ. If γ is positive (negative), the spheroid is prolate (oblate).

Let us first discuss the change of the total energy, $\tilde{E}$ in (3.22), of the liquid-droplet induced by the spheroidal deformation. This change, $\Delta\tilde{E}$, consists of the increase of the surface energy, $\Delta\tilde{E}_{surf}$, and the decrease of the Coulomb interaction energy of the charges, $\Delta\tilde{E}_{coul}$, i.e.,

$$\Delta\tilde{E} = \Delta\tilde{E}_{surf} + \Delta\tilde{E}_{coul} \, , \tag{3.42}$$

where

$$(\Delta\tilde{E}_{surf}/4\pi\sigma R_0^2 + 1) = \begin{cases} \frac{1}{2}e^{-\gamma}(1 + \frac{1}{2}e^{3\gamma}(1-e^{3\gamma})^{-1/2} \\ \quad \times\ln\{[1 + (1-e^{3\gamma})^{1/2}]/[1 - (1-e^{3\gamma})^{1/2}]\}) \\ \qquad\qquad\qquad\qquad \text{for } \gamma < 0 \\ \frac{1}{2}e^{-\gamma}[1 + e^{3\gamma}(e^{3\gamma}-1)^{-1/2}\tan^{-1}(e^{3\gamma}-1)^{1/2}] \\ \qquad\qquad\qquad\qquad \text{for } \gamma > 0 \end{cases} \tag{3.43}$$

and

$$[\Delta\tilde{E}_{coul}/(Z^2e^2/2R_0) + 1)] = \begin{cases} e^{-\gamma}(e^{-3\gamma}-1)^{-1/2}\tan^{-1}[(e^{-3\gamma}-1)^{1/2} - 1] \\ \qquad\qquad\qquad\qquad \text{for } \gamma < 0 \\ e^{-\gamma}(1-e^{-3\gamma})^{-1/2} \\ \quad \times\ln\{[1 + (1-e^{-3\gamma})^{1/2}]/[1 - (1-e^{-3\gamma})^{1/2}]\} \\ \qquad\qquad\qquad\qquad \text{for } \gamma > 0 \, . \end{cases} \tag{3.44}$$

In (3.44), Ze is the total charges distributed over the surface of a metallic liquid droplet. The derivation of (3.44) may be found in text books on electro-magnetism (for example, [3.22])]. It will be shown in Sect.3.5 that $\Delta\tilde{E}$ in (3.42) is a monotonously increasing function of $|\gamma|$.

The effective harmonic oscillator potential for the motion of an individual valence electron in the spheroid given by (3.41) is assumed to be

$$v(\mathbf{r}) = \frac{1}{2}\omega_0^2[e^{\gamma}(x^2+y^2) + e^{-2\gamma}z^2] \, , \tag{3.45}$$

whose contour of $v(\mathbf{r}) = \omega_0^2 R_0^2/2$ is the speroid given by (3.41). The one-electron energy levels in this potential are calculated analytically (Fig.3.11) as a function of γ. In this figure, we see beautiful arrays of energy gaps even at $\gamma \neq 0$. The presence of these gaps is the origin of the shell structure in non-spherical cases.

The shell correction ΔU in (3.22) for such an anisotropic harmonic oscillator model may be calculated from

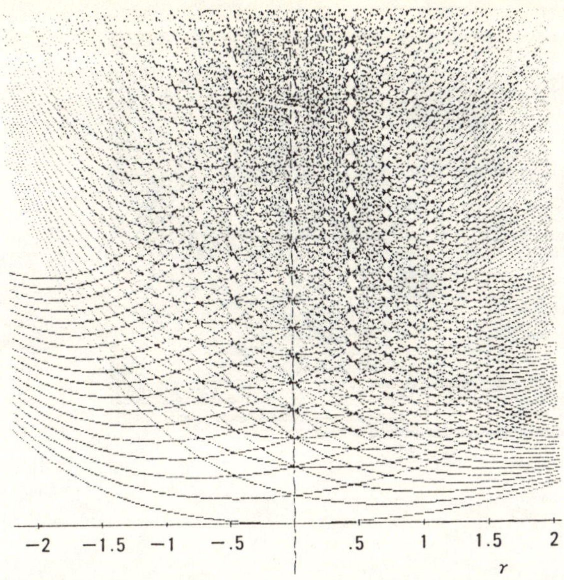

Fig.3.11. One-electron energy levels in the spheroidal harmonic oscillator potential (by A. Tamura)

$$U = \sum_{n_x, n_y, n_z} \hbar\omega_0[e^{\gamma/2}(n_x+n_y+1) + e^{-\gamma}(n_z+1/2)] \tag{3.46}$$

by using $\tilde{g}(\epsilon)$, ϵ_F/ω_0, ω_0 and $\tilde{U}$, respectively, given in (3.30, 32, 33 and 36), in each of which the second term is multiplied by the factor $(2e^{\gamma}+e^{-2\gamma})/3$, for example,

$$\tilde{U} = \frac{\omega_0}{4}\left[(3N)^{4/3} + \frac{1}{6}(3N)^{2/3}(2e^{\gamma}+e^{-2\gamma})\right]. \tag{3.47}$$

Derivation of these expressions is straightforward, so that we give here no details.

The contour map of the calculated shell correction in the two-dimensional parameter space of N and γ is depicted in Fig.3.12. The section of the map at $\gamma = 0$ shows minima at the magic numbers, N = 8, 20, 40, ···, but the minima appear at non-magic numbers in the $\gamma \neq 0$ section. In the figure, the unit of ΔU is $N^{1/3}\gamma_s^{-2}$. Since ΔU is proportional to $N^{1/3}r_s^{-2}$ for large N, as shown in (3.40), the amplitude of the oscillation of ΔU versus N is within a fixed value in the figure and the figure may be used for metals of various r_s.

By using (3.43) and Fig.3.12, total energy E_{tot} of a neutral Na_{30} cluster is calculated as exhibited in Fig.3.13, where the experimental surface ten-

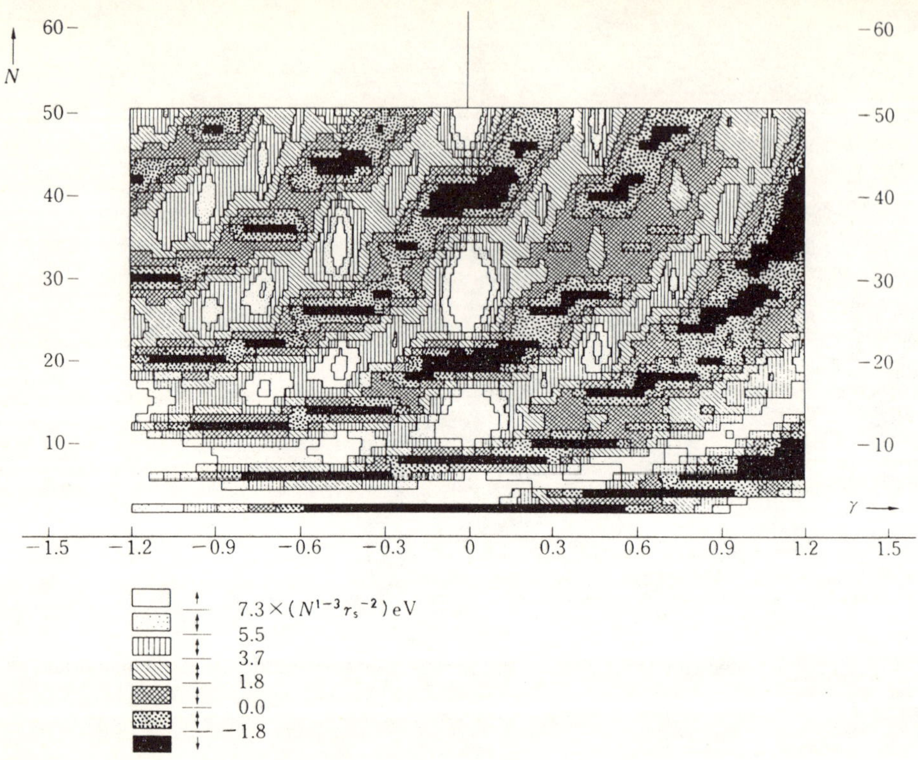

Fig.3.12. The contour map of shell correction ΔU in the spheroidal cluster of deformation parameter γ and size N (by A. Tamura)

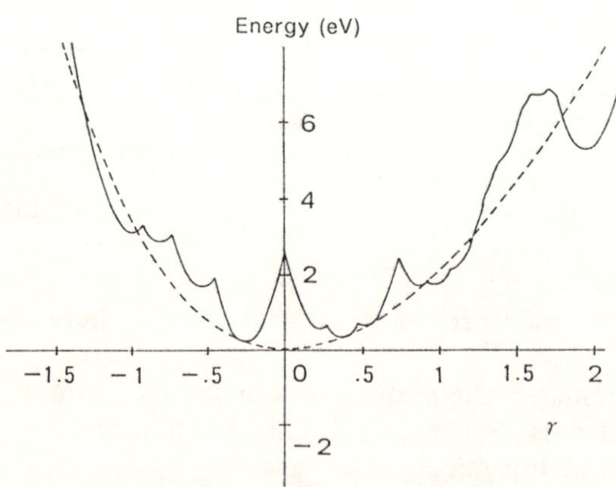

Fig.3.13. The total energy of a neutral Na_{30} cluster (*solid curve*) and the total energy of the corresponding liquid-droplet (*broken curve*) as a function of deformation parameter γ. σ = 200dyne/cm and r_s = 3.93 are assumed (by A. Tamura)

sion of liquid Na, σ = 200 dyne/cm, and the electron radius of crystalline Na, r_s = 3.93 are used. The origin of the energy scale is the energy of the corresponding liquid-droplet, $\tilde{E}$, of spherical symmetry. The figure shows

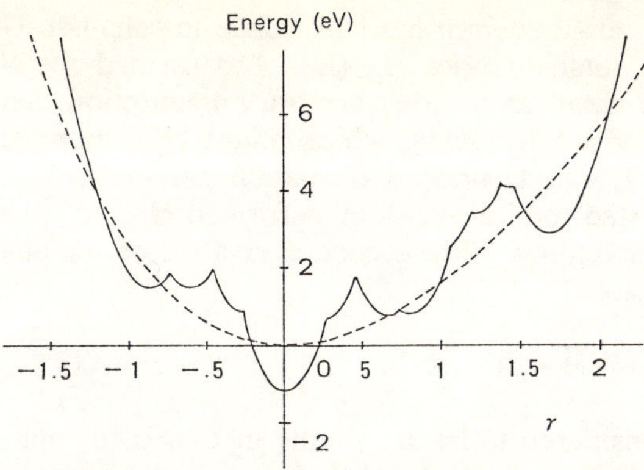

Fig.3.14. The total energy of a netural Na_{20} cluster versus γ similar to Fig.3.13 (by A. Tamura)

that a Na_{30} cluster has two stable configurations, or isomers; prolate and oblate spheroids, respectively, elongated and contracted by factor 1.5÷2.0 in the direction of the rotational axis, z. The potential barrier between these isomers is a few eV.

A similar calculation for a netral Na_{20} gives the total energy E_{tot}, as displayed in Fig.3.14. The size of this cluster, N = 20, is the shell-closing magic number for the isotropic harmonic-oscillator potential, so that the lowest minimum of E_{tot} is located at $\gamma = 0$ showing the spherical configuration to be stable.

It should be remarked that minimization of U does not give the stable deformation, as the electron-electron interaction is counted twice in U as already mentioned in Sect.3.3.1. The minimization works only when the dependence of $\tilde{E}$ and $\tilde{U}$ upon deformation parameter γ is very small.

3.4.2 Fine Structure of Mass Spectra

The effective harmonic oscillator potential in (3.45) may be supplemented by the term,

$$- \Lambda\omega_0(\ell^2 - \langle\ell^2\rangle_n) , \tag{3.48}$$

where ℓ is the angular momentum operator and $\langle\ell^2\rangle_n$ the average of the expectation values of ℓ^2 over the states with quantum number n: $\langle\ell^2\rangle_n = n(n+3)/2$. The supplemented potential is called the Nilsson potential [3.23]. The additional term is introduced to solve the degeneracy of the states with different angular momenta ℓ associated with each n in the isotropic harmonic oscillator potential. The supplement, which is equivalent to introducing anharmonicity proportional to r^4, yields energy levels closer to those of a rounded square-well potential, as indicated in the middle column of Fig. 1.6. The coefficient Λ is an adjustable parameter which is determined for the Nilsson potential to yield the energy spectrum of the jellium model. The value of Λ ranges from 0.04 to 0.08 for most sodium clusters.

By using Nilsson potential, some attempt has been made to calculate U and minimize it to obtain the total energies $E_{tot}(N)$ of deformed metal clusters of size N [3.24]. This is based on a rather arbitrary assumption that the total energy of a cluster is given by 3U/4, which might be somewhat justified if the dependence of E and U upon deformation parameter γ is very small. By using the calculated total energies of deformed clusters, the following quantity $\Delta_2(N)$ corresponding to the second derivative of the energy with respect to N is calculated:

$$\Delta_2(N) = E_{tot}(N+1) + E_{tot}(N-1) - 2E_{tot}(N) . \tag{3.49}$$

The quantity $\Delta_2(N)$ may be considered to be one of the measures for relative stability. Figure 3.15 compares the calculated $\Delta_2(N)$ with the experimental mass spectrum of sodium clusters. It is interesting to see some agreement in fine structures.

If the stable deformation is known, one is allowed to calculate the depolarization factors D_i ($i = x, y, z$) in the polarizability. On the other hand, molecular-beam deflection experiments have been performed, which measure a polarizability α averaged over all orientations:

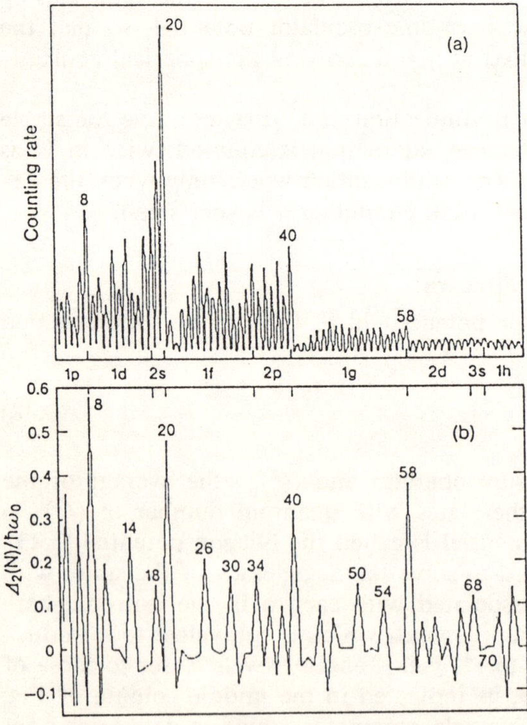

Fig.3.15. (a) The calculated relative stability $\Delta_2(N)$ of deformed clusters compared with (b) the observed Na mass spectrum [3.24]

$$\alpha = \left(\frac{1}{D_x} + \frac{1}{D_y} + \frac{1}{D_z}\right)\frac{D_0\alpha_0}{3}, \tag{3.50}$$

where α_0 and D_0 are, respectively, the polarizability and the depolarization factor of a spherical cluster. The spheroidal distortions contribute up to 8% of the total polarizability [3.14].

3.5 Fragmentation

Let us consider fragmentation of a spherical charged liquid-droplet with volume v and charge Z into two spherical charged droplets with (v_1,Z_1) and $v_2,Z_2)$. The energy gain or loss in this fragmentation is determined by the competition between the loss of the surface energy and the gain of the Coulomb interaction energy. Therefore, an important parameter directing the fragmentation is the so-called fissility parameter f [3.25] defined as

$$f = E^{(0)}_{Coul} / 2E^{(0)}_{surf}, \tag{3.51}$$

where $E^{(0)}_{Coul}$ and $E^{(0)}_{surf}$ are, respectively, the Coulomb interaction and the surface energies of a spherical charged cluster of radius R_0 as given by

$$E^{(0)}_{Coul} = (Ze)^2/2R_0, \quad E^{(0)}_{surf} - 4\pi R_0{}^2\sigma, \tag{3.52}$$

where σ is the surface tension.

Figure 3.16 shows the energy increase ΔE by the fragmentation $(v,Z{=}4) \rightarrow (v_1,Z_1){+}(v_2,Z_2)$ with r_s = 3.02 (R_0 = $r_sN^{1/3}$) and σ = 900

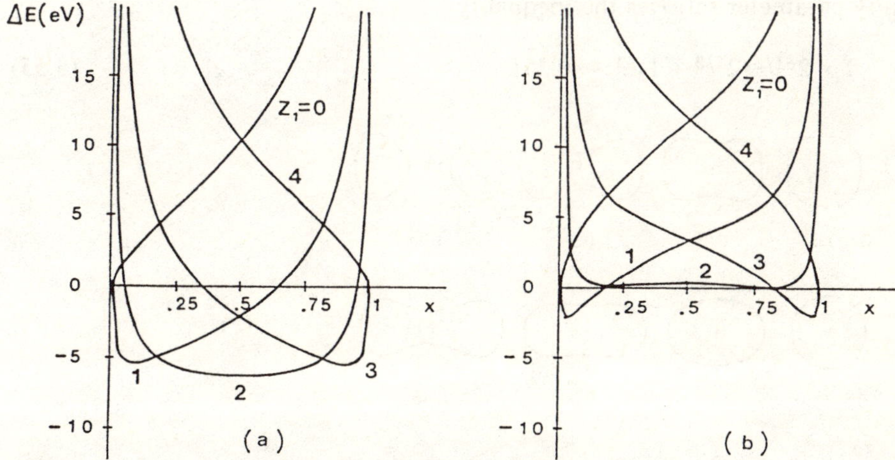

Fig.3.16. The energy increase ΔE by the fragmentation $(v,Z{=}4) \rightarrow (v_1,Z_1) + (v_2,Z_2)$ with r_s = 3.02 and σ = 900 dyne/cm for (a) N = 20; (b) N = 60. x = v_1/v and the number of the curves indicates Z_1 (by A. Tamura)

dyne/cm for (a) N = 20; (b) N = 60. The fissility parameter f for the case (a) is bigger than that of (b), which makes the energy gain ($-\Delta E$) of (a) bigger. Furthermore, we see that a large f makes the energy gain of the symmetric fission ($Z_1=Z_2$, $v_1=v_2$) prevailing.

3.5.1 Symmetric Fission of a Charged Liquid-Droplet

In what follows, we deal with only the symmetric fission of a charged metal cluster, when the charge of the cluster, Z, is large. The asymmetric fission may be dealt with, for example, following the prescription in [3.25]. The asymmetry arises from both $\tilde{E}$ and ΔU in (3.22).

To discuss the large deformation of the cluster leading to the symmetric fission, we introduce the following shape-function after the arguments on nuclear fission [3.25]

$$\frac{z^2}{R^2} + \frac{x^2 + y^2}{AR^2 + Bz^2} = 1 \, . \tag{3.53}$$

If the volume of the necked spheroid given by (3.53) is constant, $4\pi R_0^3/3$, we have the relations

$$R = CR_0 \, , \quad A = C^{-3} - B/5 \, . \tag{3.54}$$

In the case of B = 0, the shape is a prolate spheroid for $0 < A < 1$, and an oblate for $A > 1$. Non-vanishing B gives neck formation, as depicted in Fig.3.17.

Now we confine ourselves to a discussion of the classical part of the deformation energy, $\Delta\tilde{E}$. It is not difficult to show that the symmetric fission reduces the deformation energy of a charged liquid-droplet, if the fissility parameter satisfies the inequality

$$f > 2^{-1/3}(2^{1/3} + 1)^{-1} \sim 0.351 \, . \tag{3.55}$$

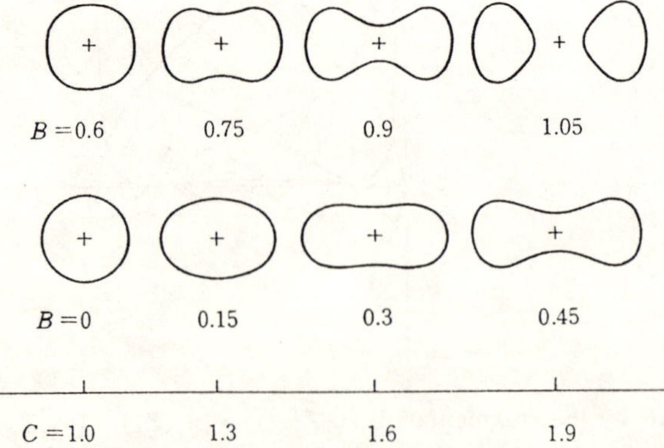

Fig.3.17. Cluster shapes and deformation parameters C and B

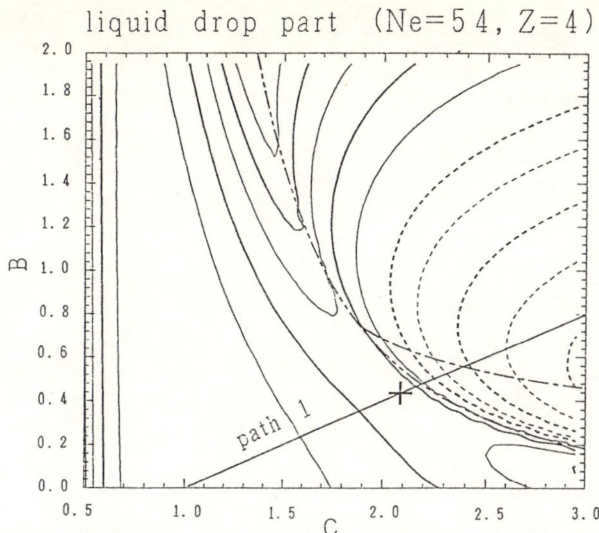

Fig.3.18. Contour map of the deformation energy $\Delta\tilde{E}$ of a charged metal droplet corresponding to Na_{58}^{4+}. (σ = 200dyn/cm, r_s = 3.93a.u.). The contours are plotted by full (for ≥0.5eV) and dotted (for ≤0.0eV) lines with the interval of 0.5 eV. Sub-contours are plotted by thin lines with an interval of 0.25 eV and the saddle point is shown by + [3.26]

It can also be shown that the spherical droplet is stable, or at least metastable, if the fissility parameter is less than unity. This fact and the condition (3.55) demonstrate that a saddle point may exist for the symmetric fission if

$$0.137 < \frac{Z^2}{N} < 0.392 \tag{3.56}$$

for parameters appropriate to sodium: We use $R_0 = r_s N^{1/3}$ with the Wigner-Seitz radius r_s = 3.93 a.u. for sodium.

The calculation of the Coulomb energy of a charged droplet of ideal conductor, where all the charges are located at the surface, needs an elaborate computer work when the shape is a necked spheroid. Figure 3.18 illustrates the contour map of the deformation energy of a charged metal droplet corresponding to Na_{58}^{4+}. In the figure, a saddle point is seen near the point, C ~ 2.1 and B ~ 0.4. The energy at the saddle point measured from that at the stable point (B=0, C=1) is around 0.6 eV.

3.5.2 Shell Correction for a Necked Spheroidal Cluster

To calculate the shell correction ΔU for our problem, we assume that an effective potential for a single electron is of an anisotropic square well with the infinite wall whose shape is given by the shape function in (3.53). The energy levels of an electron in this potential calculated numerically are

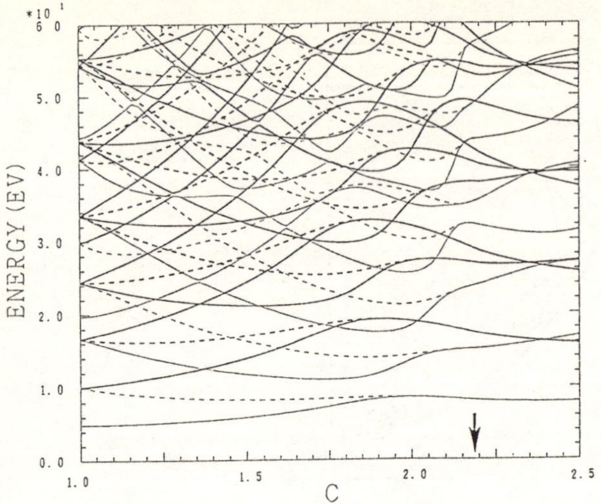

Fig. 3.19. The calculated orbital energies as a function of the deformation along the path *l* depicted in Fig.3.18. The position of split-off is indicated by an arrow. The abscissa is the values of C. Thin solid, thin dashed, thick solid and thick dashed curves are for orbitals with "even parity, m = 0", "odd parity, m = 0", "even parity, m ≠ 0" and "odd parity, m ≠ 0", respectively [3.26]

shown in Fig.3.19. The most interesting point of the figure is that the levels are again degenerate just after the cluster is split off into two pieces at C = 2.2. The reason for this degeneracy may be explained as follows. As seen in the deformation of B = 1.05 and C = 1.9 in Fig.3.17, for example, the split off pieces generally look like having trigonal symmetry when the two pieces are not far apart. Detailed examination of the potential for each piece reveals that the $\ell = 2$ term in the expansion of the potential in terms of spherical harmonics around the center of each piece vanishes on a single-dotted chain line (called "P line") in Fig.3.18 allowing the $\ell = 3$ term to predominate. Since diagonal matrix elements are vanishing of the odd-parity term of $\ell = 3$, this situation of vanishing of the $\ell = 2$ term enhances the degeneracy remarkably. We call this degeneracy *degeneracy aprés-fission.*

In order to calculate the shell correction ΔU in (3.24), we have to know the smoothed level-density function $\tilde{g}(E)$ in (3.25). This function, in the limit when electron wavelength is small compared to any characteristic dimension of the square well potential with the infinite wall, is given by the asymptotic expansion in terms of $1/k$ as

$$\tilde{g}(E) = (Vk/4\pi^2) - (S/16\pi) + (12\pi^2 k)^{-1}\tfrac{1}{2}\int d\sigma(R_1^{-1} + R_2^{-1}) , \qquad (3.57)$$

where $k^2 = E$, V is the volume inside the boundary, S the area of the boundary surface, R_1 and R_2 are the two main curvature radii at each point of the boundary, and the integral is over the boundary surface. A detailed derivation of (3.57) by using the Green's function method may be found in

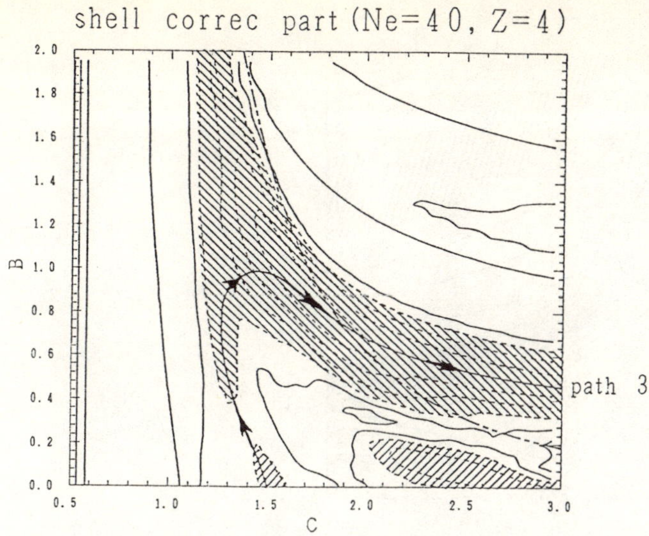

Fig.3.20. The contour map of the calculated shell correction for Na_{44}^{4+}. Contours are plotted in the same manner as in Fig.3.18. Subcontours are not plotted. In the hatched region $\Delta U \leq 0$ [3.26]

[3.27]. The first term in (3.57), which is asymptotically exact for a large volume V irrespective of its shape, is a famous theorem due to *Weyl* [3.28]. The first and the second terms give, respectively, the volume energy and the surface energy of an appropriate liquid droplet.

The shell corrections calculated from (3.18,24,25 and 57) are exhibited in Figs.3.20 and 21, respectively, for Na_{44}^{4+} and Na_{58}^{4+}. Both the Na_{44}^{4+} and its product, Na_{22}^{2+} have the magic numbers of valence electrons, but both the Na_{58}^{4+} and its product Na_{29}^{2+} non-magic numbers. The negative and positive enhancements of the shell correction are seen, respectively, in the hatched (Fig. 3.20) and shaded (Fig.3.21) regions of large deformation. These enhancements are due to the *degeneracy aprés-fission*. They are negative and positive when the products have the magic and the non-magic numbers of valence electrons, respectively: They are independent of the magicity of a mother cluster. It is important to remark that the enhancement of the shell correction is still observed along the line obtained by extrapolating the P line (the single-dotted chain line in Fig.3.18) into the unfragmented region. Along the extrapolated line near the fragmentation, the neck of the cluster is narrow enough to confine electrons on either side of the cluster.

Superposing Figs.3.18 and 21, we obtain the contour map of the total energy of Na_{58}^{4+}. However, it is found that the variation of the shell correction due to deformation predominates over that of the liquid droplet energy. Therefore, we may discuss the process of the fragmentation by using only the contour map of the shell correction. For example, we may argue that, if the magic mother cluster Na_{44}^{4+} is fragmented along the path *3*

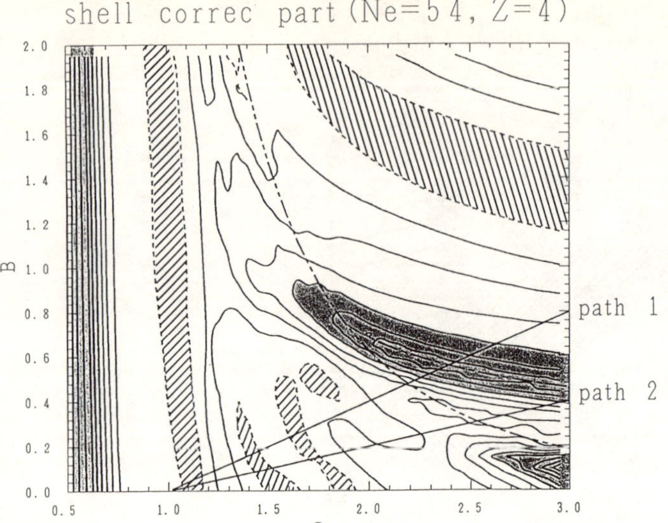

Fig.3.21. The contour map of the calculated shell correction for Na_{58}^{4+} similar to Fig.3.20. In the shaded area $\Delta U \geq 2eV$ [3.26]

shown in Fig.3.20, the activation energy is almost negligible. We may further discuss that, although the activation energy for Na_{58}^{4+} is extremely high along the path *1* indicated in Fig.3.21, it is much reduced along the path *2* in the same figure avoiding a ridge, and so on.

Finally, in Fig.3.22, we plot the energy of the shell correction at the crossing point of the path *1* and the P line, the energy of the spherical mother cluster, and their difference as a function of the number of valence electrons, N_e. Since the variation of the liquid droplet energy and the shell correction within a range of small deformations are relatively small, the difference represents the measure of barrier height along the path *1*. From this figure we notice that the barrier height is mainly determined by the shell correction of product clusters. The barrier height is reduced for the products around the magic numbers regardless of the size of the mother clusters. Therefore, we may conclude that, although the fragmentations into magic clusters are most probable, those into nearly-magic clusters are also fairly probable. Furthermore, a magic cluster, which is normally considered to be stable, is not necessarily stable if there exists a fragmentation channel into magic product clusters: an example is the case of $N_e \simeq 40$.

In Sect.3.3.2, it is pointed out that, in a specific model of the harmonic oscillator potential, the magnitude of the shell correction ΔU is proportional to $N_e^{1/3}$ while the liquid droplet energy to N_e. Generally, the magnitude of the shell correction is estimated as

$$\Delta U \sim n(\epsilon_F)\Delta E_F \, , \tag{3.58}$$

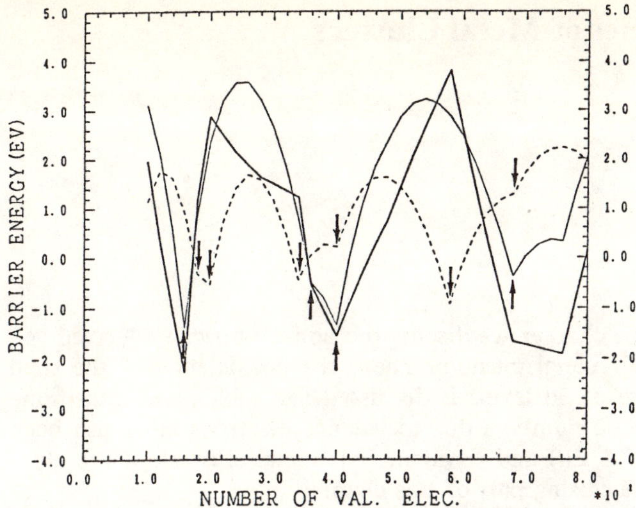

Fig.3.22. The solid, dashed and bold lines, respectively, represent ΔU at C=2.4, B=0.56 (the crossing point of the path 1 with the line P), ΔU at C=1.0, B=0.0 (the spherical mother cluster) and the differences ΔU (C=2.3, B=0.62)-ΔU (C=1.0, B=0.0), as a function of the number of the valence electrons N_e. Arrows pointing down (↓) and up (↑) represent the magic numbers for mothers and twins, respectively [3.26]

where ΔE_F and $n(E_F)$ are the mean energy gap and the degeneracy near the Fermi level, respectively. These quantities are related to the smoothed level density $\tilde{g}(E_F)$ as

$$\tilde{g}(E_F) \sim n(E_F)/\Delta E_F \; , \tag{3.59}$$

which is proportional to N_e. Then, as shown in Table 3.2, we can estimate the N_e dependence of the shell correction for three typical potentials; an irregular shaped well, a spherical well, and a spherical harmonic oscillator potential.

Table 3.2. The N_e dependence of the shell correction for various potentials

Potential	Degeneracy	$n(E_F)$	ΔE_F	ΔU
Irregularly shaped well	2 (spin)	1	N_e^{-1}	N_e^{-1}
Spherical well	2(2l+1)	$N_e^{1/3}$	$N_e^{-2/3}$	$N_e^{-1/3}$
Spherical harmonic oscillator	(n+1)(n+2)	$N_e^{2/3}$	$N_e^{-1/3}$	$N_e^{1/3}$

4. Other Properties of Metal Clusters

At the beginning of this chapter we discuss the non-empirically derived geometries of small alkali-metal clusters. Then, the coexistence of the shell structure and the group of 3d levels is discussed for noble- and transition-metal clusters. The magic numbers due to valence electrons have also been observed in divalent (Cd, Zn) and trivalent (Al) metal clusters. These clusters are discussed at the closing part of this chapter.

4.1 Nonempirical Calculation of Alkali-Metal Clusters

At the beginning of the previous chapter, to illuminate the observed shell effect, we have described a non-empirical calculation of the electronic structure of alkali-metal clusters by using the jellium model, where the geometrical shapes of the clusters, spherical in many cases, are a priori given. Later, we have discussed the theory of shell corrections, where the shapes are determined by minimizing the total energy calculated by using the simple liquid-drop model and a simple effective potential for the individual motion of a valence electron. This theory is valid, especially for the clusters whose sizes are in the range of microclusters.

A more detailed knowledge of the microscopic geometry is sometimes needed, for example to discuss catalytic reactions of the clusters. However, at the present stage it is prohibitively difficult to perform a self-consistent Hartree-Fock-type or the local-density-functional-type (Sect. 3.2.1) calculation for the clusters of sizes $N > 10$ without knowing their spatial symmetries. Such a type of calculation has been performed for the sodium clusters of $N \leq 8$ [4.1] and will be described in what follows.

In the calculation, the electronic charge and spin density of a cluster with a given geometry is calculated self-consistently within the local-spin-density approximation of the density-functional theory, as described in Sect. 3.2.1. A pseudopotential is used to treat the effect of the core electrons on the valence electrons. The equilibrium geometry may be pursued by calculating the forces acting on the constituent atoms following the theory described in the next subsection, although an accurate electronic wave function was used in [4.1] to make the additional force in (4.3) negligible. To find a true minimum, the ground state, many initial geometries are generated at random and the force calculation for each initial geometry is repeated.

4.1.1 Generalization of Hellmann-Feynman Forces

Let us denote the total energy of a cluster as $E(\mathbf{r}_1,\mathbf{R}_2,\cdots,\mathbf{R}_N\text{: E})$ which is a function of the nuclear coordinates $\mathbf{R}_j$ of constituent atoms and an external field E. According to the Hellmann-Feynman theorem, the force acting on atom μ is given by

$$\mathbf{F}_\mu^{HF} = - \partial E(\mathbf{R}_1,\mathbf{R}_2,\cdots,\mathbf{R}_N)/\partial \mathbf{R}_\mu , \tag{4.1}$$

if the electronic structure is correctly taken into account in giving the expression of the total energy. The force due to an external field is also given similarly. This condition, however, is hardly satisfied, and the Hellman-Feynman forces are well known to be extremely sensitive to the quality of the electronic wave function used in the calculation.

A famous example for showing such a situation is "a spherical neutral atom in a uniform electric field E." If the polarization of the electron cloud is ignored, the Hellmann-Feynman force is just

$$\mathbf{F}^{HF} = eZE , \tag{4.2}$$

where eZ is the nuclear charge. Obviously the correct force should be zero, and this result can be obtained if the polarization of the electron cloud is correctly taken into account.

Recently, it has been shown that the addition of the following force $\mathbf{F}_\mu'$ to $\mathbf{F}_\mu^{HF}$ in (4.1) gives a better expression of the force acting on the μth atom, when the electronic wave functions, as given by Linear Combinations of Atomic Orbitals (LCAO), are approximate [4.2]

$$\mathbf{F}_\mu' = - \left[\sum_i^{occ} \int d\mathbf{r} (d\psi_i/d\mathbf{R}_\mu)(h - \epsilon_i)\psi_i + \text{c.c.} \right] , \tag{4.3}$$

where ψ_i is the approximate electronic wave function of the ith orbital, and h is the Hartree-Fock-type Hamiltonian acting on a single electron derived from an approximate electron distribution. We see that $\mathbf{F}_\mu'$ is vanishing if ψ_i satisfies the equation $(h-\epsilon_i)\psi_i = 0$. Otherwise $\mathbf{F}_\mu'$ gives a contribution of the same order of magnitude as that of $\mathbf{F}_\mu^{HF}$.

4.1.2 Geometries Calculated

In this subsection, the geometries of sodium clusters calculated in [4.1] are shown without mentioning details of the calculation. For details the reader is referred to the original paper. The calculated equilibrium geometries of neutral and ionized clusters with N $\leq$ 8 are exhibited in Figs.4.1-3. The geometry of Na_3 is an isosceles triangle (C_{2v} symmetry) and that of Na_3^+ is an equilateral triangle (D_{3h} symmetry). Both Na_4 and Na_4^+ have a planar rhombic geometry (D_{2h} symmetry). The Na_5 cluster has a planar geometry

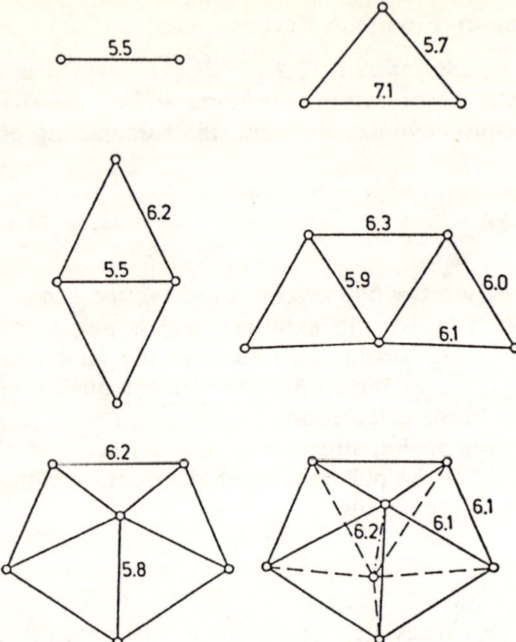

Fig.4.1. Equilibrium geometries of Na_N clusters with $N \leq 7$. The internuclear distances are given in atomic units [4.1]

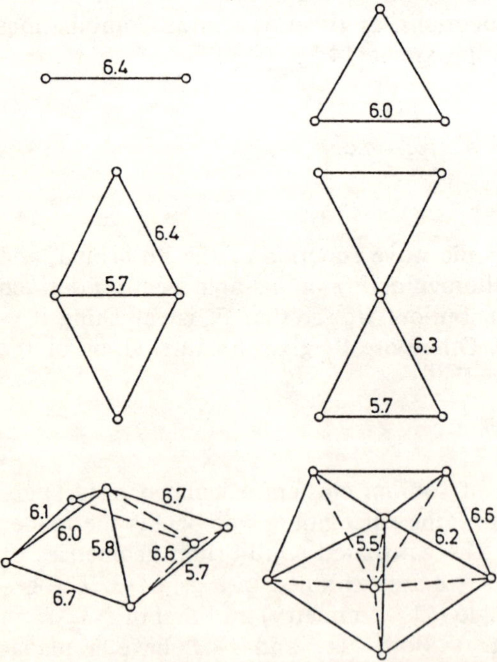

Fig.4.2. Equilibrium geometries of Na_N^+ clusters with $N \leq 7$, similar to Fig.4.1 [4.1]

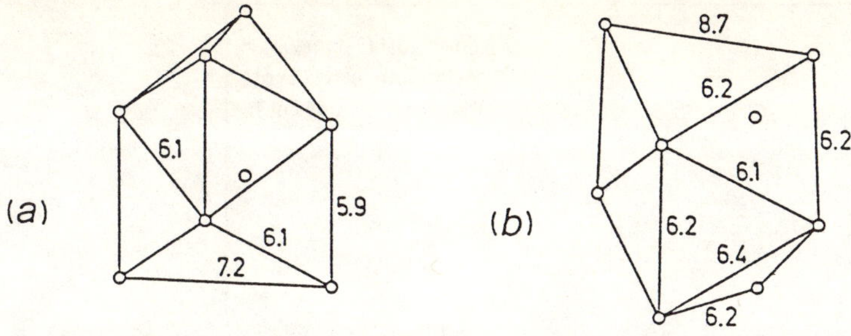

Fig.4.3. Equilibrium geometries of (a) Na_8 and (b) Na_8^+ [4.1]

with an almost trapezoidal shape (C_{2v} symmetry) formed of three slightly distorted equilateral triangles. The Na_5^+ cluster is composed of two isosceles triangles with a common apex (D_2 symmetry). They can rotate almost freely around the axis passing through the apex atom and the middle points of the bases of the triangles. The special case of a planar geometry (D_{2h} symmetry) is depicted in Fig.4.2. Furthermore, the bending modes of one triangle with respect to the other are very soft.

The equilibrium geometry of Na_6 is a flattened pentagonal pyramid (C_{5v} symmetry). The energy surface of this cluster has a local minimum lying only 0.04 eV above the lowest minimum. This isomer, which is not shown here, has a planar structure formed by a central equilateral triangle surrounded by three isosceles triangles (D_{3h} symmetry). The Na_6^+ has a very low symmetry (C_s), as illustrated in Fig.4.2, although it may be regarded as a distortion of a bcc-like structure. The energy surface of Na_6^+ has a local minimum with D_{2h} symmetry 0.03 eV above the absolute minimum. This isomer is not shown here.

Both the Na_7 and the Na_7^+ clusters have a pentagonal bipyramidal structure (D_{5h} symmetry). The Na_7 has an almost regular structure where all the edges have approximately the same length, but the Na_7^+ is a flattened bipyramid.

As displayed in Fig.4.3, the Na_8 and Na_8^+ have rather compact structures with D_{2d} and D_{2v} symmetries, respectively. The Na_8^+ has a slightly distorted bcc-like structure and can be obtained from the Na_8 geometry through a relatively large distortion. The Na_{13} and Na_{13}^+ are also studied by assuming a few symmetrical distortions of the cubo-octahedron and of the icosahedron, in contrast with the calculation of the equilibrium geometries of the other clusters where no a priori geometry is assumed. The results of Na_{13} and Na_{13}^+ are not shown here.

The results mentioned in this subsection can be summarized as follows: (i) Sodium clusters with five atoms or less have a
planar structure built from distorted equilateral triangles, which is the typical pattern of close packing in two dimensions; (ii) the clusters with six, seven and eight atoms have three-dimensional structures and a regularity of the interatomic distances starts to appear, although the clusters with six electrons or less are flattened with respect to an ideal geometry.

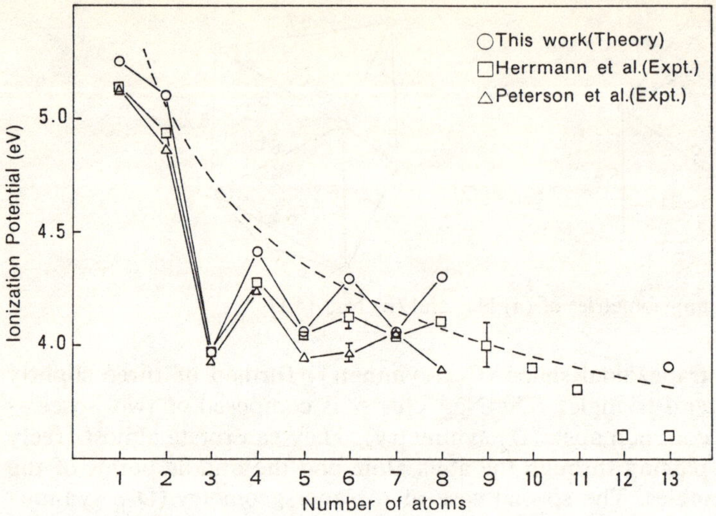

Fig.4.4. Ionization potentials of Na_N clusters. Experimental error bars are shown only for Na_6 and Na_g. The broken line is the result of a simple electrostatic model [4.1,3,4]

4.1.3 Comparison with Experiments

The appearance potential of Na_N clusters in a molecular beam have been measured by photoionization followed by mass spectroscopy [4.3,4]. The appearance potential, which is the threshold of the energy of the ionizing particle that produces a measurable ion signal, may normally be considered to be equal to the adiabatic ionization potentials. In Fig.4.4, the calculated adiabatic ionization potentials of Na_N are compared with the experimental values. In the figure the results of a simple electrostatic model are also depicted [4.5-7]. The calculated ionization potentials, in particular their trend, are in better agreement with the results of *Herrmann* et al. [4.3].

The electron-spin resonance of alkali clusters trapped in a frozen inert-gas matrix have been reported [4.8-10]. The clusters are assigned to trimers and septamers. The spectra of pentamers have not been identified. Interpretation of the observed hyperfine structure of the spectra gives the electronic spin density at the nuclear sites, which is often expressed as a fraction of the atomic value called the isotropic spin population. At low temperature the spectrum of Na_3 has large and equal spin densities at two nuclear sites and a smaller density at a third nucleus. The calculated isotropic spin population is in good agreement with the experimental one, as can be seen from Table 4.1. In particular, the agreement of the sign should be emphasized, which is determined by comparing with the low-temperature spectra the high-temperature spectra (T>20K) where the rapid dynamical Jahn-Jeller pseudorotation gives the same spin density at all three nuclei.

Table 4.1. Experimental and calculated isotropic spin populations. The multiplicity is indicated in parentheses [4.1]

	Experiment	Calculated
Na	1.0	1.0
Na_3	0.47 (2×)	0.48 (2×)
	-0.07	-0.04
Na_5		0.19 (2×)
		0.17 (2×)
		0.06
Na_7	0.37 (2×)	0.32 (2×)
	±0.02 (5×)	-0.02 (5×)

The spin resonance of Na_7 shows large and equal spin densities at two nuclei and smaller densities at five other nuclei. The results are consistent with the theoretically predicted geometry of a pentagonal bipyramid (Fig. 4.1). The calculated isotropic spin populations are also in good agreement with the experimental ones (Table 4.1).

4.1.4 Nature of Binding

In Sects.3.1,2 it was mentioned that alkali and noble-metal clusters show the shell structure which arises from the motion of a valence electron in a spherically symmetric effective potential: the motion is specified by the orbitals, 1s, 1p, 1d, 2s, ···, in the order of increasing energy and this order is rather insensitive to a detailed form of the effective potential. It was also mentioned in Sect.3.2.3 that degeneracy of these energy levels is removed by a non-spherical perturbation, but the splitting is small as compared with the energy separation of the levels of the spherical symmetry. Normally, one may expect a non-spherical distortion of the degenerate system by the Jahn-Teller effect when the energy levels of the spherical symmetry are not filled. Therefore, the following electron configurations are expected for the clusters of size N;

$$\begin{aligned}
N = 2:&\quad (1s)^2\ ,\\
N = 3:&\quad (1s)^2(1p_x)\ ,\\
N = 4:&\quad (1s)^2(1p_x)^2\ ,\\
N = 5:&\quad (1s)^2(1p_x)^2(1p_y)\ ,\\
N = 6:&\quad (1s)^2(1p_x)^2(1p_y)^2\ ,\\
N = 7:&\quad (1s)^2(1p_x)^2(1p_y)^2(1p_z)\ ,\\
N = 8:&\quad (1s)^2(1p_x)^2(1p_y)^2(1p_z)^2\ ,
\end{aligned}$$

where the $1p_x$ is a non-degenerate, split component of the 1p level with the lowest energy, and the $1p_y$ is the non-degenerate component orthogonal to the $1p_x$, and so on. The axes x, y and z, are not fixed in space, but the systems with the electron configuration of the partially or completely filled $1p_x$ and $1p_y$ and the empty $1p_z$ ($N \leq 6$) are expected to have the planar distribution of valence electrons. This expectation agrees with the results (Figs. 4.1,2) of the non-empirical calculation mentioned in Sect.4.1.1: The Na_N ($N = 3,4,5$) and the Na_N^+ ($N = 4,5$) have planar geometries; the Na_6, Na_6^+ and Na_7^+ are flattened ones.

The systems with the electron configuration of the partially or completely filled $1p_z$ are expected to have the three-dimensional distribution of valence electrons. This agrees with the results (Figs.4.1,3) of the calculation that the Na_N ($N = 7,8$) and Na_8^+ have three-dimensional geometries.

The considerations just mentioned above seem to show that the electronic structure of valence electrons is the principal factor for determining the general feature of the equilibrium geometry of the clusters and the main feature of the electronic structure of valence electrons are rather insensitive to the geometrical structure. This conclusion is clearly related to the delocalized free-electron-like nature of valence electrons of the sodium clusters. Now we wonder how far this conclusion is applicable to other metal clusters.

4.2 Electronic Structure of Noble-Metal Clusters

Noble metals have one delocalized valence electron per atom, which is interacting with the atomic d-electrons to a certain degree. It is our present understanding that in noble-metal clusters, these valence electrons are free-electron-like to such an extent to show the shell structure, as described in Sect.3.1. The purpose of the present section is to clarify the interrelation between the shell structure of valence electrons of noble-metal clusters and the energy bands of noble-metals by citing a specific example of small copper clusters.

4.2.1 Energy Levels of Copper Clusters

Assuming the geometries of small copper clusters, a local-density-functional-type calculation by using the so-called Slater's Discrete-Variational exhcange-alpha (DV-Xα) method has been performed to derive energy levels of the valence electrons and the atomic d-electrons [4.11]. The calculation is made for six Cu_N with $N = 4, 6, 8, 13, 14$ and 19, whose atomic structures are depicted in Fig.4.5. The symmetry is tetrahedral for $N = 4$ and 8, and octahedral for $N = 6, 13, 14$ and 19. The bulk crystals of the fcc structure contain these clusters except $N = 8$. For all these clusters, the nearest neighbor distance is assumed to be that of the bulk, 4.71 a.u.

Figure 4.6 exhibits the calculated energy levels. The length of each level indicates the 3d population by Mulliken's charge analysis. It is clearly seen in the figure that a group of the energy levels of large 3d population, simply called the 3d band, is located in the energy range of $E \sim -0.2$ to

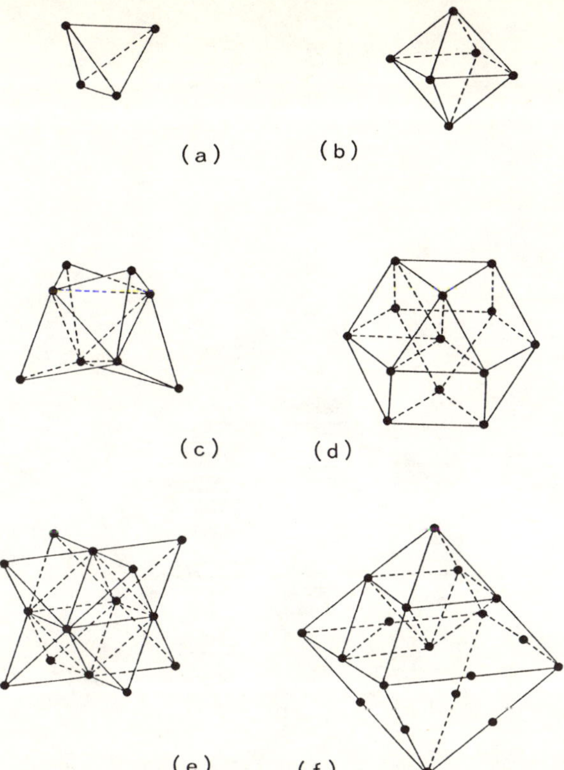

Fig.4.5. Atomic structure of clusters: **(a)** N = 4 tetrahedron, **(b)** N = 6 octahedron, **(c)** N = 8 tetrahedron multi-twinned, **(d)** N = 13 cuboctahedron, **(e)** N = 14 combined cluster of octahedron and cube, **(f)** N = 19 combined cluster of cuboctahedron and octahedron [4.11]

-0.3. The energy levels of small 3d population are distributed below and above the 3d band. The lowest one of these denoted as A_{1g} (or A_1 for N = 4,8) corresponds to the 1s level of the shell structure for an electron in the spherically symmetric square well, as shown in Fig.1.6, the next T_{1u} (or T_2 for N = 4,8) of small 3d population, which is above the 3d band for N = 4, 6 and 8 but below it for N = 13, 14 and 19, to the 1p, the E_g and T_{2g} (or E and T_2 for N = 4,8), which are above the 3d band for N = 4, 6, 8, 13 and 14 but below it for N = 19, to the 1d, the higher A_{1g} (or A, for N = 4,8) which are above the 3d band for N = 4, 6, 8, 13 and 14 but below it for N = 19, to the 2s, and so on. It is quite interesting to see the result that the number of the energy levels of the shell structure below the 3d band increases as the cluster size increases: the E_g and T_{2g} (the 1d shell structure) first appear below the 3d for Cu_{19}.

In Fig.4.6, the occupied levels are depicted by thick lines and the unoccupied ones by broken lines. The highest occupied level, the Fermi level, is a level of the shell structure above the 3d band: the T_2 (1p) for N =

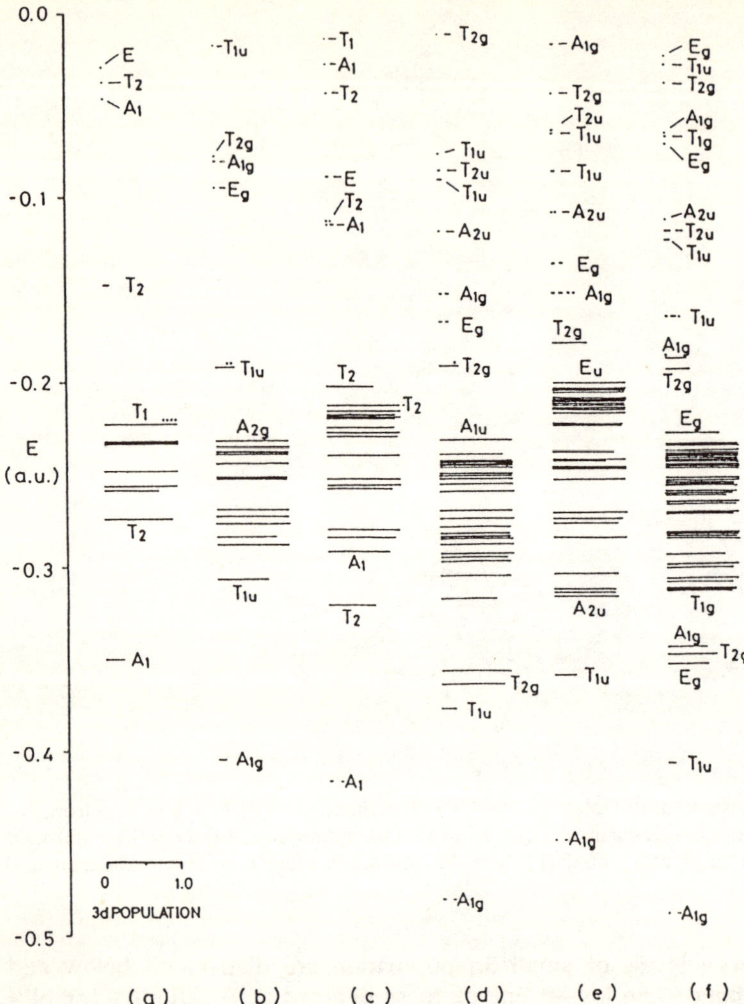

Fig.4.6. The calculated energy levels of (a) Cu_4, (b) Cu_6, (c) Cu_8, (d) Cu_{13}, (e) Cu_{14} and (f) Cu_{19}, as depicted in Fig.4.5. The length of the horizontal bar for each level indicates the 3d population. Full and dashed lines represent occupied and unoccupied levels, respectively [4.11]

4, the T_{1u}(1p) for N = 6, the T_2(1p) for N = 8, the T_{2g}(1d) for N = 13 and 14 with the E_g(1d) unoccupied, the A_{1g}(2s) for N = 19 with the T_{2g}(1d) and E_g(1d) occupied. This situation agrees with that of the shell model without taking into account the 3d levels. The level of the shell structure near the 3d band has a small amount of the 3d character. It would be interesting to detect such a 3d character by some experiment.

Since the cluster geometries are assumed a priori, the absolute positions of the energy levels, consequently the Fermi levels, in Fig.4.6 are rather approximate. Nevertheless, the theoretical prediction in Sect.3.3.2 is seen to be approximately realized in Fig.4.6 that the Fermi level approaches a fixed value when the cluster size increases. This implies that the number of energy levels of the shell structure increases as the cluster size increases below the 3d band if the position of the 3d band relative to the Fermi level and its width do not change so much, as really seen in Fig.4.6.

More elaborate, all-electron SCF calculation with a careful choice of basis sets has been performed for Cu_N (N = 2, 3, 4, 5, 6) with fixed geometries [4.12]. This calculation shows that the Fermi level is 1.6÷2.2 eV above the 3d levels (3d band) for Cu_5 and Cu_6, although it is ~2.1 eV for the bulk. The results suggest that Cu_5 and Cu_6 already have a property of the Cu bulk as far as the photoionization is concerned. Only a qualitative difference from that of the bulk is quantization of the s band according the boundary condition, i.e., appearance of the shell structure. The equilibrium bond length with a fixed symmetry, binding energy, and the bonding charge distribution for Cu_2, Cu_4, Cu_{13} and Cu_{79} clusters are investigated by using the self-consistent local-density functional theory [4.13]. One-electron energy levels, however, have not been investigated in detail in this study.

4.2.2 Photoelectron Spectra of a Copper Cluster

The ultraviolet photoelectron spectra of mass-selected negatively-charged copper clusters in a form of cluster beam are measured at a photon energy of 4.66 eV. Size of the clusters are from 6 through 41 atoms [4.14]. The clusters are produced by laser vaporization of a copper disc mounted on the side of a pulsed supersonic nozzle. Clusters are formed in the near-sonic flow of the helium carrier gas. The clusters are irradiated by a laser pulse to be negatively charged, and finally subjected to free supersonic expansion into the main vacuum chamber of a cluster beam apparatus. Neon gas (2%) is added to the helium carrier gas to improve the cooling of the clusters. The internal temperature of the clusters is estimated to be below 300K.

In Fig.4.7 the photoelectron spectra measured for Cu_N^- (N = 6÷41) are depicted. On each spectrum an arrow indicates an estimated energy of the photodetachment threshold, i.e., the electron affinity. For the clusters of even N, a horizontal bar shows an estimate of the energy gap between the Highest Occupied Molecular Orbital (HOMO) and the Lowest Unoccupied Molecular Orbital (LUMO). In the clusters of even N, an extra electron is assumed to occupy the LUMO of the corresponding neutral cluster, while the next least-bound electron to occupy the HOMO of the same neutral cluster. The HOMO-LUMO gap can then be extracted from the photoelectron spectrum of each even-N cluster as the energy difference between the lowest energy peak in the binding energy spectrum and the onset of the next feature.

Figure 4.8 compares the measured electron affinities with the results of a simple treatment of the shell-model by using Nilsson potential with ellipsoidal distortions, as described in Sect.3.4.2. Since a high energy for the HOMO of a negatively charged cluster implies a low electron affinity, the

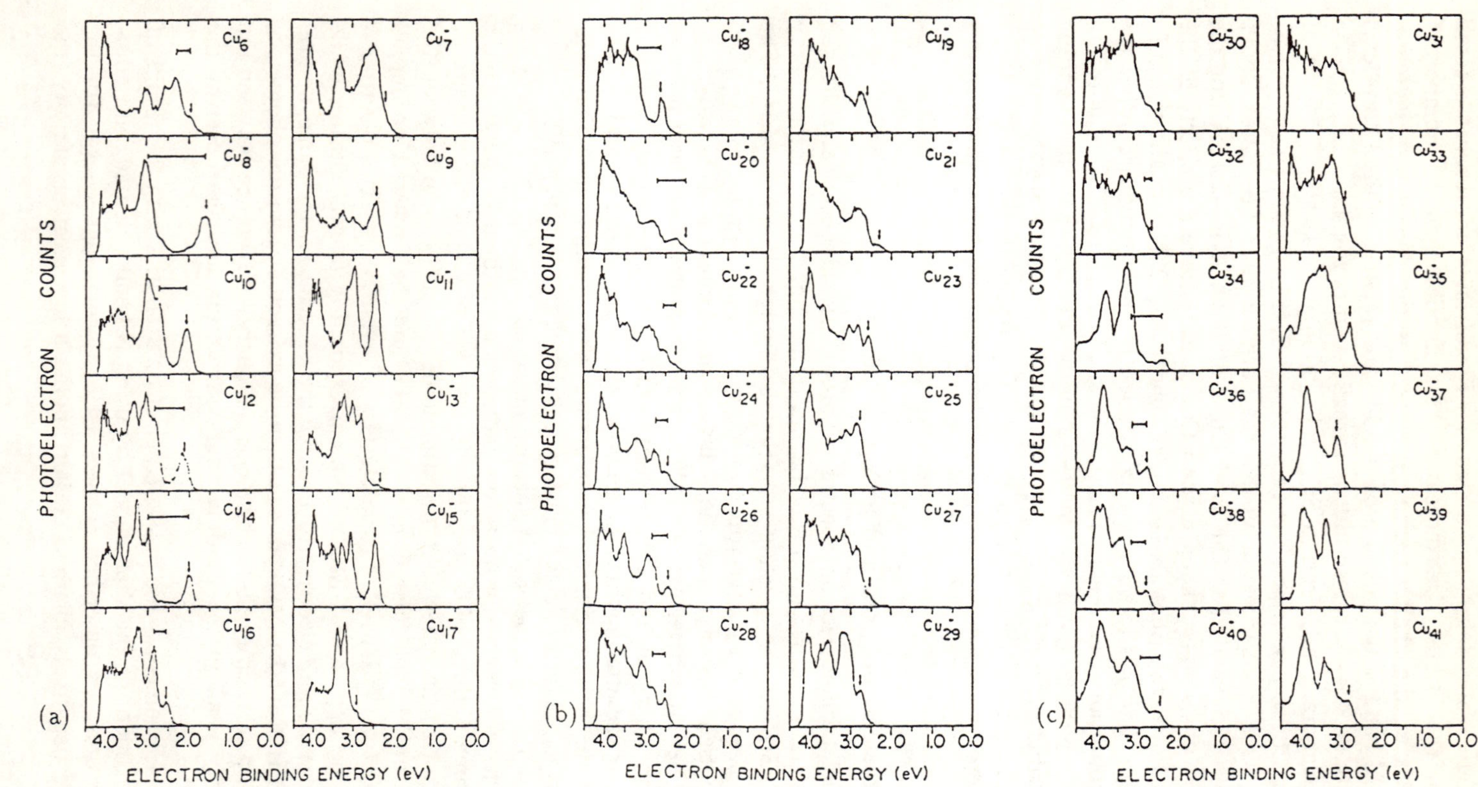

Fig.4.7a–c. Ultraviolet photoelectron spectra of mass-selected negative copper clusters generated by the 4.66 eV photon excitation. Arrows indicate the electron affinities. For the even clusters the estimated HOMO-LUMO gap of the corresponding neutral cluster is marked by a horizontal bar. The data in the range of the binding energies larger than 4.0 eV are unreliable [4.14]

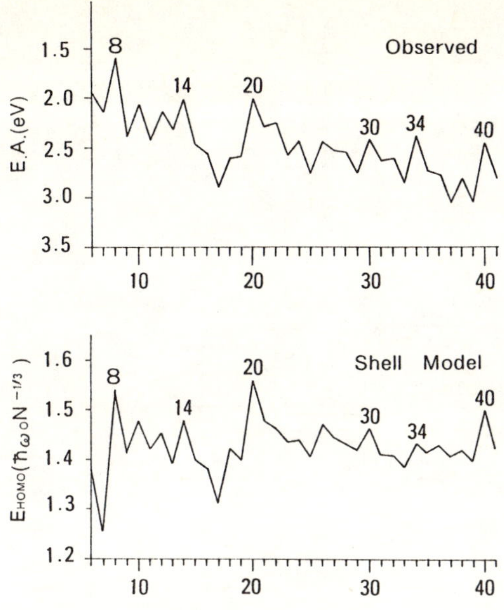

Fig.4.8. The measured vertical electron affinities of Cu_N^- as a function of N (top) compared with the highest energy levels of ellipsoidally distorted shell model for N+1 electrons (bottom) [4.14]

electron affinity is plotted on an inverted scale. It is clearly seen that the clusters with spherical shell closings for the neutral ones (N = 8,20,40) and subshell closings (N = 14,34) have abnormally low affinities, as predicted by the shell model. The measured HOMO-LUMO gaps also exhibit anomalies at these magic numbers of N.

As mentioned in Sect.4.2.1, theoretical calculations predict the onset of the 3d band roughly 2 eV below the Fermi level. However, it seems to be premature from the experimental points of view to ascribe the observed feature below 3 eV in binding energy to the 3d band.

4.3 Electronic Structure of Transition-Metal Clusters

4.3.1 Energy Levels of Nickel Clusters

The total number of the s valence electrons and the d-electrons of a neutral nickel atom is ten. Accordingly, the effective number of d-holes is equal to that of s-electrons as far as we are confined to the s- and d-states. By assuming the geometries of clusters, as shown in Fig.4.5, a spin-polarized DV-Xα calculation has been performed to derive energy levels of the valence s-electrons and the d-electrons [4.15].

Figure 4.9 displays the calculated spin-polarized energy levels of Ni clusters. The length of the horizontal line for each level indicates the 3d population. Full and dashed lines represent occupied and unoccupied levels, respectively. It is seen in the figure that a group of energy levels of large 3d

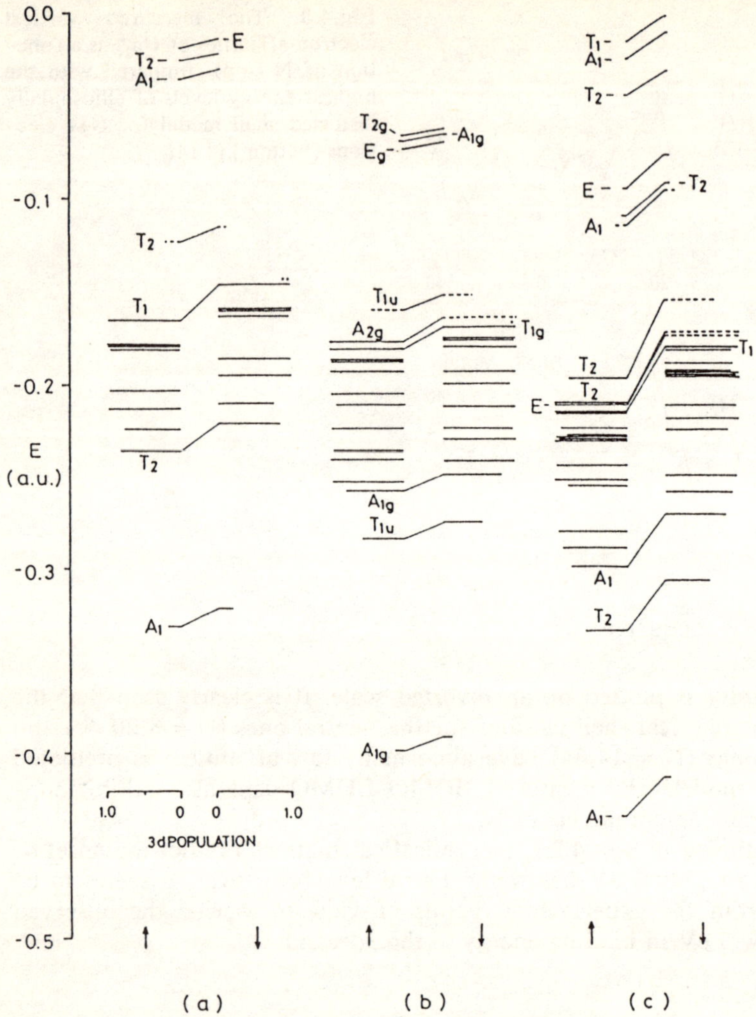

Fig.4.9a–c. For the caption see the opposite page

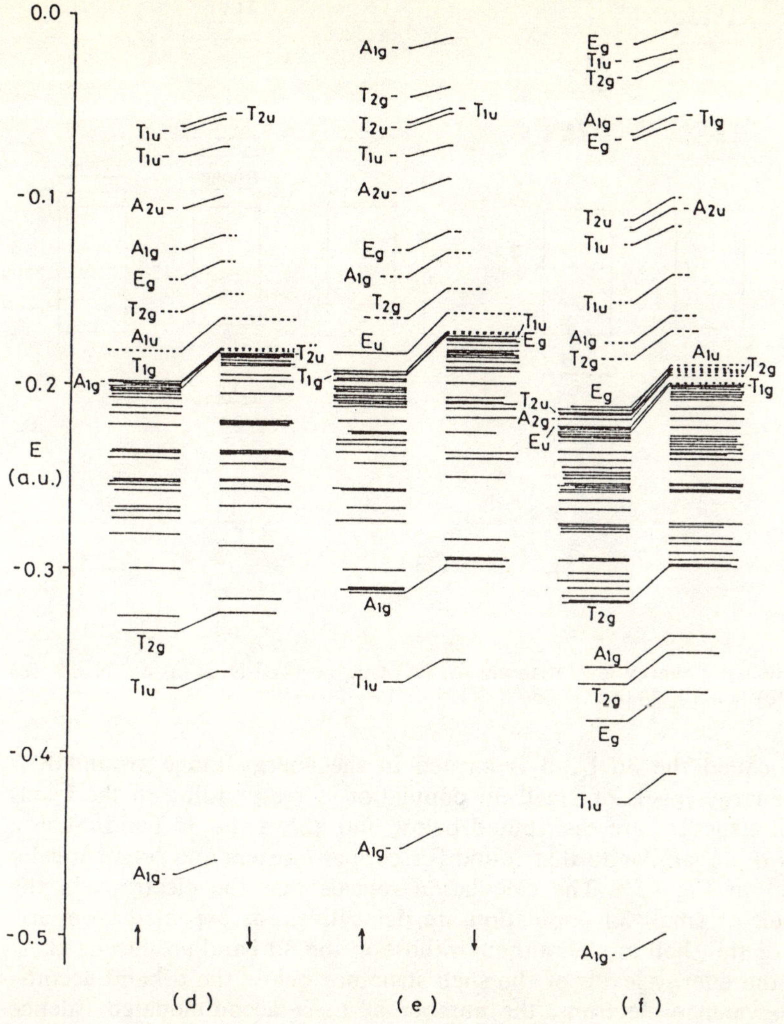

Fig.4.9. The calculated spin-polarized energy levels of (a) Ni_4, (b) Ni_6, (c) Ni_8, (d) Ni_{13}, (e) Ni_{14} and (f) Ni_{19} whose atomic structures are shown in Fig.4.5. The length of the horizontal line for each level indicates the 3d population. Full and dashed lines represent occupied and unoccupied levels, respectively [4.15]

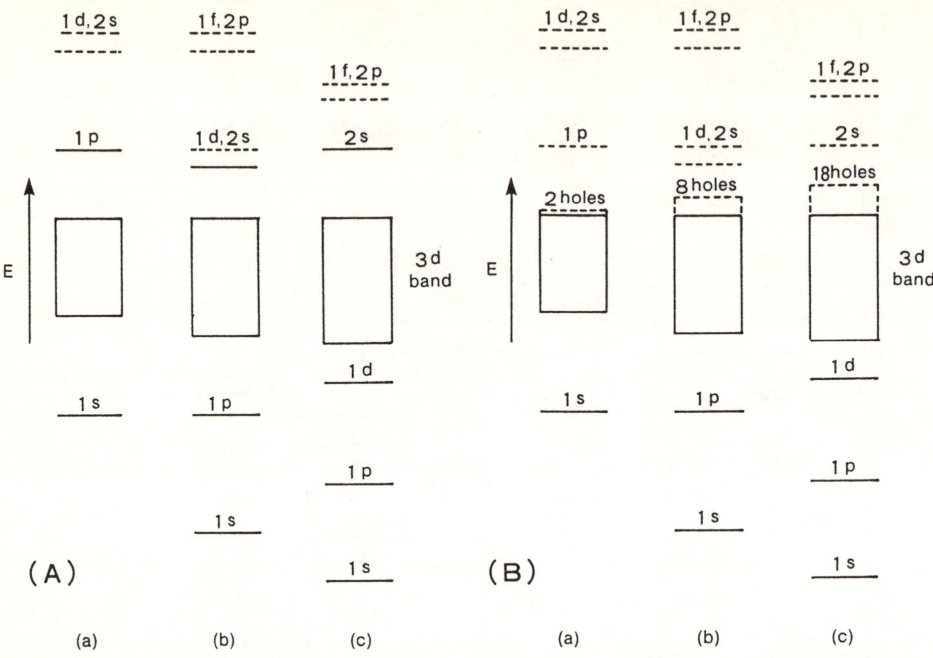

Fig.4.10. Schematic energy level diagram for (**A**) Cu_N and (**B**) Ni_N. (a) $4 < N < 8$, (b) $8 < N < 18$, (c) $N = 19, 20$ [4.15]

population called the 3d band is located in the energy range around E ~ -0.2. The energy levels of small 3d population corresponding to the levels of the shell structure are distributed below and above the 3d band. Such a situation is quite similar to that found for copper clusters and schematically summarized in Fig.4.10. The calculation reveals that the electrons in the energy levels of small 3d population are delocalized, as expected from applicability of the shell model, although those in the 3d band are localized.

Since the energy levels of the shell structure below the d band accommodate the valence electrons, the number of these accommodated valence electrons gives that of d-holes in nickel clusters (Fig.4.10) if the shell levels have no 3d population. The situation, however, is not as simple as this: the 1d shell level of Ni_{19} just below the d band has more than 50% 3d population, that reduces the effective number of d-holes. The calculation gives the number of unpaired electrons (Table 4.2). It is most remarkable that the number shows a stepwise increase as N increases. Such a discontinuous change in the number may somewhat be smoothed out by the mixing of the 3d band and the shell level: the mixing is enhanced when the energy separation between them happens to be small by some reason or due to the splitting of the shell levels by distortion. The relative positions of the d band and the shell levels of the valence s electrons depend upon the method of the calculation adopted [4.16-18].

Table 4.2. The calculated number of unpaired electrons. The number can be either 8 or 6 for Ni_{13}, due to the accidental degeneracy of the up-spin A_{1u} LUMO and the down-spin HOMO, as shown in Fig.4.9 [4.15]

N	4	6	8	13	14	19
The approximate number of unpaired electrons	2	2	8	8 (6)	8	12

4.3.2 Stern-Gerlach's Experiment on Iron Clusters

Magnetic properties of Fe_N with N = 2÷17 as well as those of Fe_NO and Fe_NO_2 with N = 2÷7 have been measured by the use of Stern-Gerlach-type experimental arrangement [4.19]. A schematic diagram of the pulsed cluster-beam apparatus and the geometrical arrangement for magnetic deflection of metal clusters are exhibited in Fig.4.11. The magnetic properties of clusters are examined in two ways. The spatial deflection is measured by spatially translating the ionizing laser beam across the cluster beam at a fixed magnetic field strength. Alternatively, the ionizing laser is set at a fixed position relative to the cluster beam, and the field strength is varied to deflect individual magnetic substates through the detection region.

Referring to the coordinate system shown in Fig.4.11, the deflection in the z direction of a species, i, with magnetic moment μ_i and mass m_i is given as

$$d_i = \mu_i \frac{\partial H}{\partial z} L^2 \frac{1 + 2D/L}{2m_i v_{xi}^2} \tag{4.4}$$

where L is the length of the magnet, $\partial H/\partial z$ is the gradient of the magnetic field, D is the distance from the end of the magnet to the detection region, and v_{xi} is the velocity of the species through the magnet along the x axis. Since $\partial H/\partial t$ can be determined by measuring d_i and v_{xi} for the species i with known μ_i and m_i, the magnetic moments μ_j for other species j with known m_j can determined by measuring d_j and v_{xj}.

The measurements seem to be still at a preliminary stage, but indicate that all the clusters have the magnetic moments which increase as the cluster sizes increase. The results seem to be consistent with a simple physical picture in which very strong spin ordering is achieved. The experiment is so qualitative that no step-wise increase of the magnetic moment can be examined, as discussed in the previous subsection for nickel clusters.

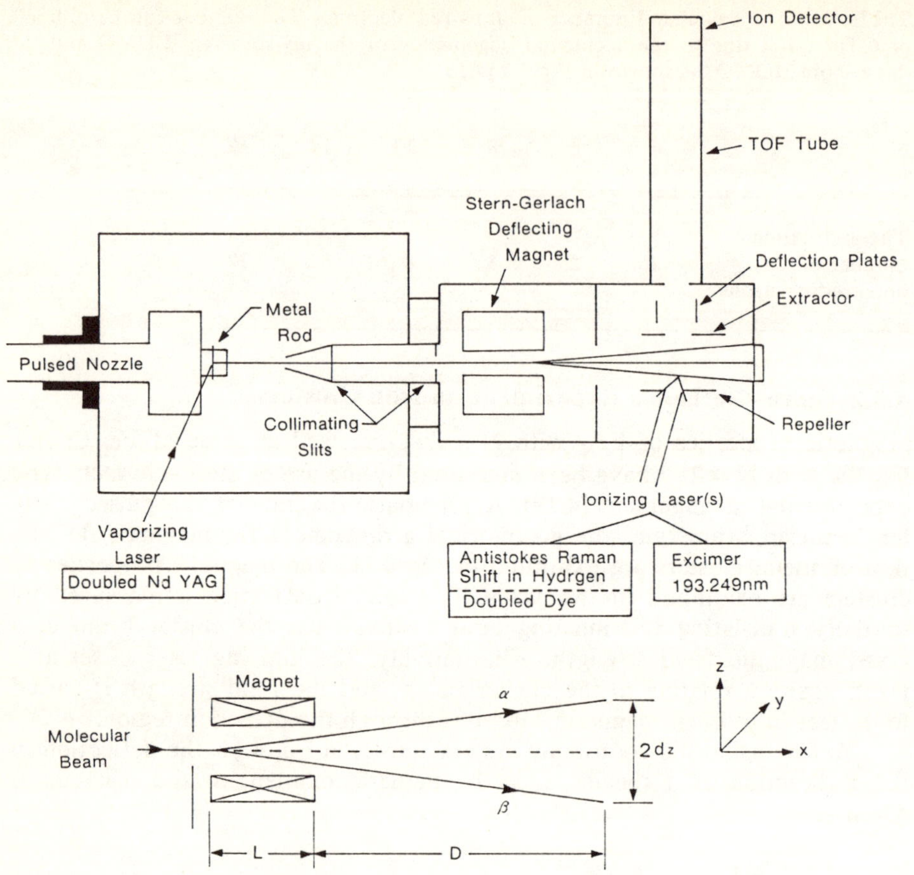

Fig.4.11. Schematic of the pulsed molecular-beam apparatus and the geometrical arrangement for magnetic deflection of the metal clusters [4.19]

4.4 Divalent-Metal Clusters

The mass spectra of singly-charged Zn and Cd cluster have been measured [4.20]. In these cases, the constituent atoms have two valence electrons, so that the total number of valence electrons of the singly charged clusters is always odd. The observed mass spectra of the positive and negative Zn clusters are displayed in Fig.4.12. The trace of the negative clusters is shifted by one cluster size towards the right-hand side. The ion intensities of the positive clusters up to N = 6 and of the negative clusters up to N = 24 are not determined because of high background noise. The traces of the observed two mass spectra are quite similar: high ion intensities are observed at N = 10, 18, 20, 28, 30, 32, 35, 40, 41, 46, 47, 54, 57, 60 and 69 for the positive clusters, and at N = 27, 29, 31, 34, 40, 45, 46, 52, 56, 60, 61 and 68 for the negative clusters. Similar traces of mass spectra are obtained for Cd clusters.

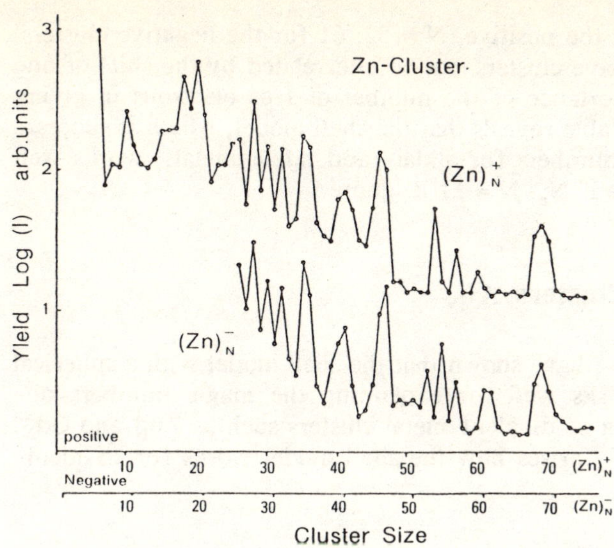

Fig.4.12. Size distributions of singly-charged zinc clusters [4.20]

Table 4.3. Magic number N of $(Zn_N)^{\pm}$, the number of free electrons N_e, and the shell closing number $\bar{N}$ [4.20]

N (positive ion)	N (negative ion)	Ne	$\bar{N}$
10		19	20
18		35	34
20		39	40
28	27	55	-
30	29	59	58
32	31	63	-
-	-	-	68
35	34	69	68
40	-	79	-
41	40	81	-
46	45	91	92
-	52	105	106
54	54	107, 109	106
57	56	113	112
60	60	120, 121	-
-	61	123	-
69	68	137	138

Table 4.3 gives the cluster size N of zinc clusters, at which peaks of the ion intensity are observed, the number of free electrons N_e, and the shell closing number $\bar{N}$ for the square well potential with a round edge corresponding to the middle column of Fig.7.6. The fact that almost all the

peaks, except N = 40 for the positive, N = 52, 61 for the negative clusters, for the positive and negative clusters can be interrelated by the shift of one cluster size indicates importance of the number of free electrons in giving the observed peaks. The table reveals that the shell model, which is successful in giving the magic numbers for alklai- and noble-metals, works well for divalent metal clusters if $N_e - \overline{N} = \pm 1$ is ignored.

4.5 Trivalent-Metal Clusters, Al_N

In the previous section, we have shown that the shell model with a spherical square-well potential works well for explaining the magic numbers observed in the mass spectra of divalent-metal clusters such as Zn_N^+ and Cd_N^+. Then, a question naturally arises how the shell model works for trivalent-metal clusters such as Al_N.

4.5.1 Observed Properties

A mass spectrum of the aluminum cluster ions generated by pulsed laser vaporization of an aluminum rod in a continuous flow of helium buffer gas is depicted in Fig.4.13 [4.21]. The measured spectrum varies with source conditions, laser power and ion beam focusing voltages. Intense peaks are seen in the spectrum for Al_7^+ and Al_{14}^+. Similar features have also been observed in a different experiment [4.22].

Collision-induced dissociation of Al_N^+ (N = 3÷26) is measured with a centre of mass collision energy of 5.25 eV and a pressure of approximately 2.0 mTorr of argon in a gas cell [4.21]. Only four products, Al_{N-1}^+, Al_{N-2}^+,

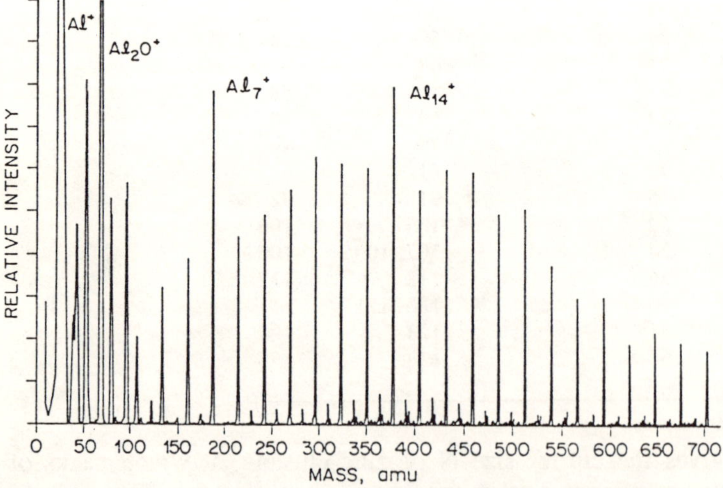

Fig.4.13. Mass spectrum of aluminum cluster ions. The small peaks are mainly due to trace contaminants. Some are due to double charged aluminum clusters, which first appear for Al_{13}^{2+} between Al_6^+ and Al_7^+: Al_{15}^{2+} is absent [4.21]

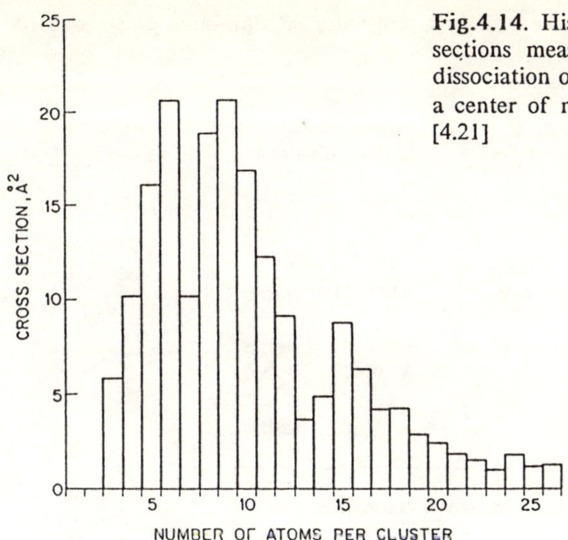

Fig.4.14. Histogram showing the total cross sections measured for the collision-induced dissociation of Al_N^+ by argon (2.0 mTorr) with a center of mass collision energy of 5.25 eV [4.21]

Al_{N-3}^+ and Al^+, are observed with any significant intensity. It is observed that Al^+ is the main product of the smaller ($N \leq 14$) clusters. Between Al_{14}^+ and Al_{15}^+ a sharp transition occurs and the main product becomes Al_{N-1}^+ for clusters with fifteen and more atoms.

The cross sections for collision-induced dissociation are derived from

$$\sigma = -\frac{1}{n\ell}\ln\left(\frac{I_R}{I_R + \Sigma I_p}\right), \tag{4.5}$$

where I_R and I_p are the reactant and product ion intensities, respectively, n the collision gas density, and ℓ the gas cell length. The cross sections measured with approximately 2.0 mTorr of argon in the gas cell are indicated in Fig.4.14. Since the measured cross sections vary with the gas cell pressure, the values in Fig.4.14 are not true cross sections at the zero pressured limit. Besides a broad maximum at Al_6^+-Al_9^+, there are sharp drops in the cross sections at Al_7^+, Al_{13}^+, Al_{14}^+, and Al_{23}^+.

The product branching ratios of the collision-induced dissociation contain information about the ionization potentials of the clusters. Analyzing the branching ratios in an approximate way [4.21], one obtains the ionization potentials as a function of the cluster size. In the figure, we see that the ionization potentials initially rise with cluster size. The maximum occurs at Al_6, and they fall with the increase of cluster size. The ionization potential of Al_7 is substantially lower than its neighbors and a sharp drop is seen at Al_{14} where the ionization potential drops below that of the atom.

Now we are in a position to discuss the relationship of the observed properties of aluminum clusters mentioned above to the electronic shell model, as summarized in Table 4.4. The high ionization potential of Al_6

Table 4.4. The relationship of the observed properties of aluminum clusters to the electronic shell model [4.21]

Shell	Shell-closing number	Properties of neutral clusters		Properties of ionized clusters	
1p	8				
1d	18	$Al_6(18)$[a]	high IP		
2s	20	$Al_7(21)$	low IP	$Al_7^+(20)$	stable
1f	34				
2p	40	$Al_{13}(39)$	high IP	$Al_{13}^+(38)$	stable
		$Al_{14}(42)$	low IP	$Al_{14}^+(41)$	stable
1g	58	$Al_{20}(60)$	low IP		
2d	68			$Al_{23}^+(68)$	stable

[a] The number of valence electrons is shown in parentheses

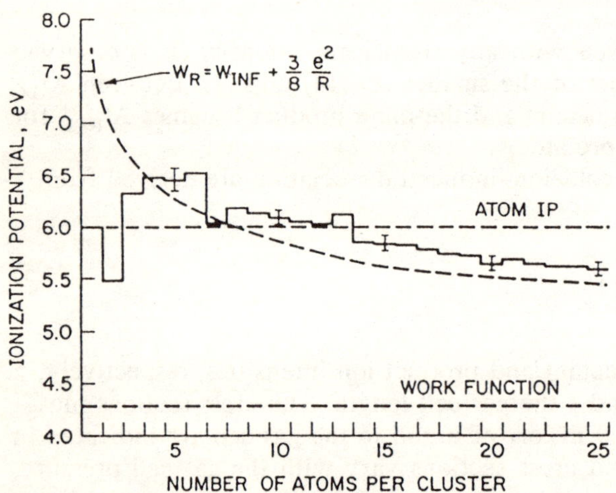

Fig.4.15. Ionization potentials of the aluminum clusters derived from the measured product branching ratios. W_R and W_{INF} are, respectively, the work functions of a conducting sphere of radius R and the bulk [4.21]

and Al_{13} (Fig.4.15) could be understood as they have closed and nearly closed shells, respectively. The low ionization potentials of Al_7 and Al_{14} could be understood as they have, respectively, one and two excess electrons over closed shells and these electrons are easily removed. The enhanced stability of Al_7^+, as shown in Figs.4.13,14, could be related to the closing of the 2s shell. The stability of Al_{13}^+ and Al_{14}^+ could be related to the approximate closing of the 2p shell. In Fig.4.15, one sees that the ionization potential of Al_{20} is marginally lower than its neighbors. This may be related to the closing of the 1g shell. In Fig.4.14 the collision-induced dis-

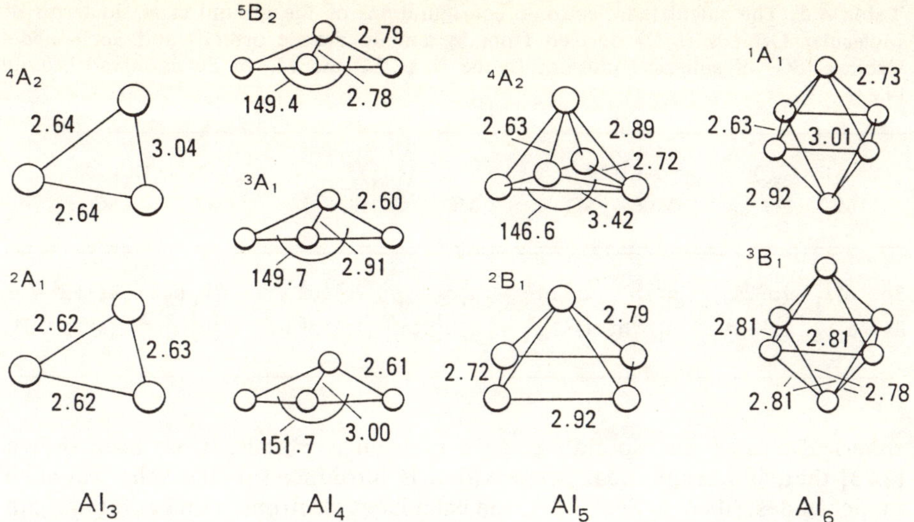

Fig.4.16. The calculated geometries of Al_3-Al_6 for the lowest states of each spin multiplicity [4.23]

sociation cross section of Al_{23}^+ is significantly lower than its neighbors. This may be related to the closing of the 2d shell. As discussed so far, some of the experimental data could be accounted for on the basis of the shell model. However, one sees no real evidence for the closing of the 1f shell and the evidence for the 1g shell closing is rather weak.

4.5.2 Nonempirical Calculation of Al_2-Al_6

The first-principles total-energy calculations for optimized structures of Al_N ($N = 2 \div 6$) are performed by using a pseudopotential [4.23]. The calculated geometries of Al_3-Al_6 in the lowest states within each spin manifold are depicted in Fig.4.16. The geometries are three-dimensional starting with Al_4. This behavior is quite different from that of small alkali clusters, as described in Sect.4.1.2. According to the calculation, the states of the lowest spin multiplicity are the lowest in energy except Al_6, where the states of $S = 0$ and 1 are almost degenerate. Within the set of low-spin clusters, the geometric tendency is towards fcc substructures, but not necessarily to close-packed geometries. For example, Al_4 is closer to planar than tetrahedral, while Al_5 is a square pyramid rather than the more closed-packed trigonal bypyramid.

Clusters of Al_2-Al_6 are too small to be the clusters to which the shell model is applicable. Nevertheless, it is of some interest to discuss the interrelationship between the results of the non-empirical calculation and the shell model. Table 4.5 summarizes the electronic configurations of the ground states for Al_3-Al_6 obtained from the non-empirical calculation. The configurations are given in terms of molecular orbitals derived from the 3s and 3p atomic orbitals of an aluminum atom and shell-model orbitals of a

Table 4.5. The calculated electronic configurations of the ground states in terms of Molecular Orbitals (MO) derived from 3ş and 3p atomic orbitals and Shell-model Orbitals (SO) of spherical clusters. In the 2s of SO the $1d_0$ of SO is mixed heavily [4.23]

	Al_3 (C_{2v})		Al_4 (C_{2v})		Al_5 (C_{2v})		Al_6 (D_2)	
	MO	SO	MO	SO	MO	SO	MO	SO
3p	$b_1^2a_1$	$1p^2 2s$	$a_2^2a_1^2$	$2s^2 1d^2$	$a_1^4b_2$	$2s^2 1d^3$	$a_1^4b_2b_3$	$1s^2 1d^4$
3s	$a_1^4b_2^2$	$1s^2 1p^4$	$a_1^4b_1^2b_2^2$	$1s^2 1p^6$	$a_1^4b_1^2b_2^2a_2^2$	$1s^2 1p^6 1d^2$	$a_1^4b_1^4b_2^2b_3^2$	$1s^2 1p^6 1d^4$

spherical cluster corresponding to the molecular orbitals. It has been shown [4.23] that, if nonspherical perturbation is introduced to the spherical shell model as described in Sect. 3.2.3, the calculated electronic configurations can be explained by the shell model with a nonspherical cavity.

In order to see the quantitative nature of the non-empirical calculation cited above and the experimental work, the experimental ionization potentials of Al_2-Al_6 are compared with the calculated one in Table 4.6. Agreement between the values of different work looks fair. As for the experimental values of the total cross-section of the collision-induced dissociation, however, some discrepancy has been found between different experiments on small aluminum clusters [4.24].

Table 4.6. A comparison of the experimental and calculated ionization potentials in eV for Al_2-Al_6

Reference	Al_2	Al_3	Al_4	Al_5	Al_6
Experiment					
[4.21]	5.6	6.3	6.4	6.5	6.7
[4.22]	6.0÷6.4	6.4÷6.5	6.5÷7.9	6.4÷6.5	6.0÷6.4
[4.24]	5.2	5.9	6.1	6.0	6.2
Theory					
[4.23][a]	6.0	6.3	6.5	5.6	6.6

[a] Derived from product branching ratios by using the method of *Jarrold* [4.21]

5. Semiconductor Clusters

Semiconductor clusters are the microclusters whose constituent atoms are those of semiconductor crystals such as C, Si, Ge, etc. This does not necessarily mean that semiconductor clusters are semiconducting. It is well known that the picture of covalent bonding is applicable to semiconductor crystals, where valence electrons are localized around the constituent atoms forming directional bonds. Then, it is interesting to ask the question, what kind of picture is applicable to understand the chemical bonding of semiconductor clusters. Is the picture entirely different from that for metal clusters where valence electrons are delocalized, making an individual motion?

5.1 Carbon Clusters

5.1.1 Mass Spectra

Laser vaporization of a substrate within the throat of a pulsed nozzle in the atmosphere of helium-carrier gas is used to generate a supersonic beam of carbon clusters [5.1]. The neutral cluster beam is photoionized by an ultraviolet laser and probed with a time-of-flight mass analizer. By using graphite as the substrate, carbon clusters C_N^+ for $N = 1 \div 190$ are produced. A typical mass spectrum of carbon clusters thus produced exhibits a distinctly bimodal size distribution, as depicted in Fig.5.1.

In the size range of $N < 30$, an ion signal is seen for each C_N, although there are distinct periodic alterations in ion intensity with a period of $\Delta N = 4$. In the size range of $N > 40$, ion signals are observed only for the clusters of even N. From the observed dependence of ion signals upon the ionizing laser power it is believed that the observed mass spectrum, as depicted in Fig.5.1, reflects the neutral cluster distribution in the cluster beam.

When the density of helium gas is increased to maximize cluster thermalization and cluster-cluster reactions before the passage through the nozzle, a remarkable enhancement of the C_{60}^+ ion signal is observed as compared with signals of the other cluster ions [5.2]: The mass spectrum shows essentially a single line of the C_{60}^+ signal in the range of $N > 40$. Experimental determination of the geometrical structure of C_{60} has not been successful yet.[1] Nevertheless, many people have no serious objection to the

[1] Quite recently, crystallization of C_{60} has been demonstrated successfully. The crystal structure was studied by X rays [5.12].

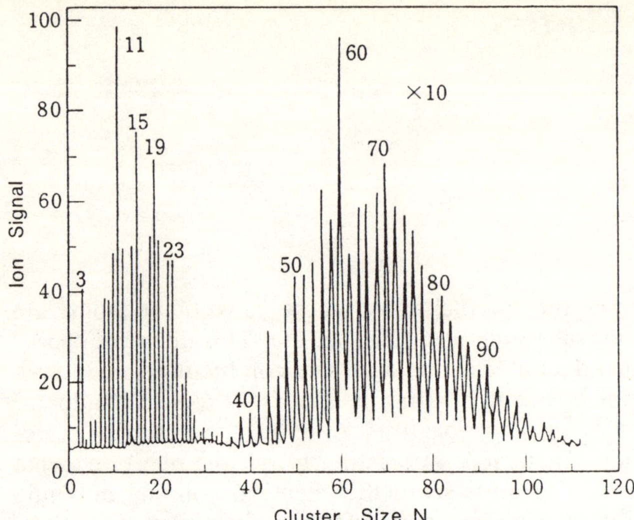

Fig.5.1. A typical photoionization time-of-flight mass spectrum of carbon clusters. The gain for C_N^+ (N>40) is increased by a factor of 10. The signal intensities of the two distributions (N<30 and N>0) cannot be compared directly, as the experimental conditions are different [5.1]

Fig.5.2. A soccerball with the pentagonal symmetry. Twelve pentagonal black patches are isolated by hexagonal white patches

following speculation on the geometry: sixty carbon atoms are located at five corners of twelve pentagonal patches on the surface of a soccerball, as illustrated in Fig.5.2. The structure has the icosahedral symmetry. The inner and outer surfaces are covered with a sea of π electrons. The diameter of the ball is about 7Å, providing an inner cavity capable of accommodating a variety of atoms.

5.1.2 Nonempirical Calculation

The structure and energies of small carbon clusters, C_2-C_{10}, are studied by performing accurate ab initio calculations [5.3]. For the determination of cluster geometries, several possible geometrical arrangments including linear, cyclic, and some three-dimensional forms are assumed. For each ar-

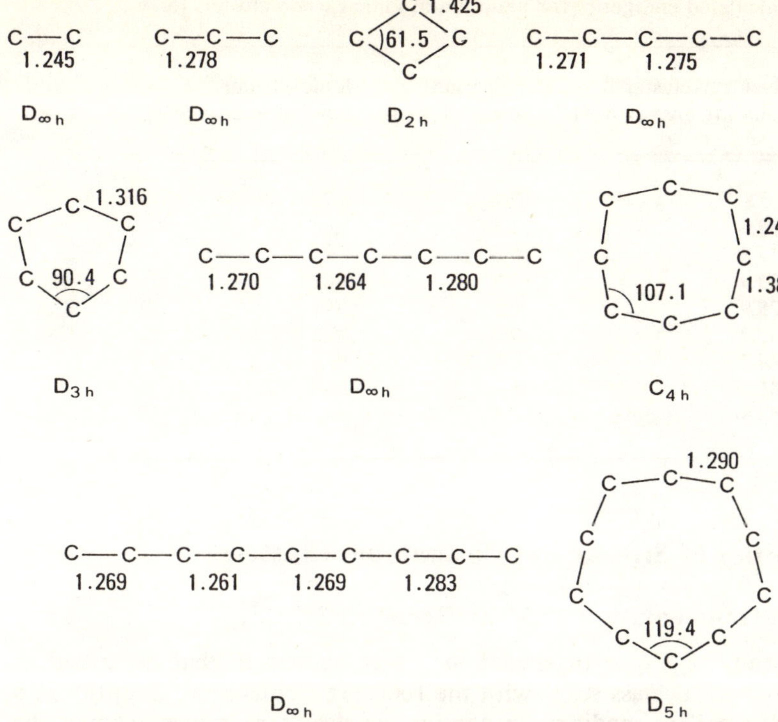

Fig.5.3. The calculated ground-state geometries of C_2-C_{10}. Bond lengths are shown in Ångstroms and bond angles in degrees. For C_4, C_6 and C_8, low-lying linear structures exist which are close in energy to the cyclic forms [5.3]

rangement the geometry is completely optimized with the given symmetry constraints. The results are depicted in Fig.5.3.

The geometry optimizations reveal several interesting aspects. All the clusters are found to have either linear or monocyclic ground states involving multiple bonding. The linear structures are lowest in energy for the clusters of odd N, and monocyclic for those of even N. The cyclic structures benefit from the additional bonding resulting from ring closure. However, the energy gained from such bond formation has to be weighted against the strain energy which may become appreciable in the more compact cyclic structures. This balance between two opposing factors causes many of the even-numbered clusters to have monocyclic ground states.

The calculated energies of these carbon clusters are summarized in Table 5.1. Detailed comparison of the calculated binding energies of neutral carbon clusters with the corresponding experimental values of small clusters suggests that approximately 90% of the binding energy is reproduced by these calculations. However, it is still difficult to extend the calculations of this accuracy to larger carbon clusters and discuss their stabilities, as revealed in the observed mass spectra (Fig.5.1).

Table 5.1. Calculated energetics for neutral and ionic carbon clusters [5.3]

Cluster	Neutral-cluster binding energy [eV]	Ionization potential [eV]	Ionic-cluster binding energy [eV]
C_2	5.8	12.1	4.7
C_3	12.7	11.4	12.3
C_4	17.3	10.5	17.8
C_5	23.9	10.7	24.2
C_6	28.9	9.8	30.1
C_7	35.2	10.0	36.2
C_8	40.3	9.2	42.1
C_9	46.4	9.4	48.0
C_{10}	53.9		

5.2 Stabilities of Silicon and Germanium Clusters

5.2.1 Photofragmentation of Mass-Resolved Si_2^+-Si_{12}^+

A cluster beam of Si_N^+ is produced in a way similar to that described in Sect.5.1.1. A typical mass scan, with the focusing elements set to optimize a collection of small to medium clusters at the detector, is displayed in the lower portion of Fig.5.4 [5.4]. A set of "switch-out" electrodes selects only clusters with the mass of interest to proceed further. The selected cluster is slowed down by the deceleration grids and exposed to an intense beam of 267, 355 or 532 nm pulsed laser radiation. The resulting charged photofragments are reaccelerated and dispersed in the second half of the time-of-flight region. The inset of Fig.5.4 shows the mass spectrum of the fragments when Si_{12}^+ is irradiated by 266 nm light.

The relative cross sections for the photofragmentation of Si_N^+ into Si_M^+ (N = 2÷12) at 532 nm are given in Table 5.2. These values for a fixed N are largely independent of whether the fragmentation-laser photon energy is 2.3, 3.5 or 4.7 eV, although the absolute cross sections are considerably larger at 4.7 eV than at 2.3 eV. Table 5.2 reveals that, when the ionic cluster breaks up, positive charge remains predominantly on the larger fragment and, for breakup of the clusters of N = 7÷11, the fragment Si_6^+ is unusually prominent.

The relative total cross sections are illustrated in Fig.5.5 for photofragmentation of Si_N^+ (N = 2÷11) at 532 nm. These values are the cross sections of Si_N^+ summed over all fragmentation channels, and are measured by recording the saturation in the depletion of the original beam of Si_N^+ as a function of laser intensity. For the measurement, the apparatus is programmed to record simultaneously the various Si_N^+ cross sections relative to that of Si_6^+, thus avoiding most of the systematic errors associated with ab-

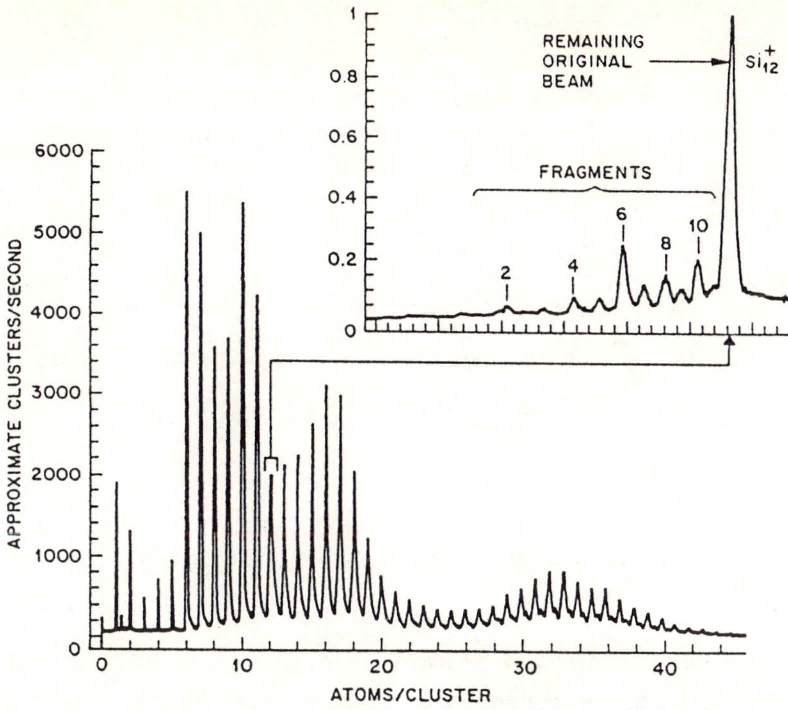

Fig.5.4. The observed mass spectrum of silicon cluster ions. The inserted figure shows the mass spectrum of product cluster ions when a Si_{12}^+ cluster is photofragmented [5.4]

Table 5.2. Branching ratios for the fragmentation of Si_N^+ into Si_M^+ [5.4]

N / M	2	3	4	5	6	7	8	9	10	11	12
1	1.00	0.25	0.17	0.05	0.00	0.01	0.00	0.00	0.00	0.00	0.00
2		0.75	0.18	0.05	0.04	0.02	0.01	0.00	0.00	0.00	0.02
3			0.65	0.08	0.05	0.01	0.01	0.02	0.01	0.02	0.00
4				0.82	0.21	0.03	0.07	0.06	0.09	0.09	0.05
5					0.70	0.11	0.07	0.08	0.03	0.13	0.05
6						0.82	0.28	0.39	0.64	0.29	0.05
7							0.55	0.19	0.12	0.36	0.08
8								0.26	0.07	0.04	0.12
9									0.04	0.03	0.09
10										0.04	0.42
11											0.13

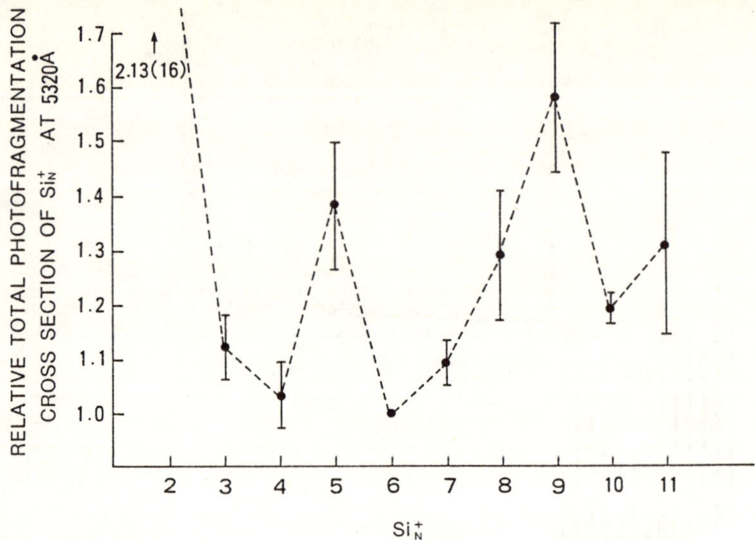

Fig.5.5. Relative total photofragmentation cross sections of Si_N^+ at 532 nm. Si_6^+ is used as a reference [5.4]

solute cross-section measurements. Figure 5.5 exhibits that Si_4^+, Si_6^+ and Si_{10}^+ have relatively small total photofragmentation cross sections. This fact, together with the remarkable abundance of Si_6^+ and Si_{10}^+ observed in the mass spectrum of Fig.5.4, indicates that Si_6^+ and Si_{10}^+ are quite stable.

5.2.2 Photofragmentation of Larger Silicon and Germanium Clusters

Photofragmentation of Si_N^+ with size N up to 60 and Ge_N^+ with N up to 50 are studied by using Q-switched Nd:YAG, KrF and ArF excimer laser lines [5.5]. Similar studies have also been performed on the negative cluster ions. The fragmentations depend upon the photoexcitation intensities and size of the clusters. The results are not so simple, but we shall summarize some systematic tendencies found for positive cluster ions in what follows.

For Si_N^+ with $N > 11$, only $N = 6 \div 11$ positive ion fragments are recorded even at high intensity excitations. In the range of $N = 12 \div 24$, the following symmetric fission processes are often observed;

$$Si_{12}^+ \rightarrow Si_6^+ + Si_6^+ ,$$
$$Si_{14}^+ \rightarrow Si_7^+ + Si_7^+ ,$$
$$Si_{20}^+ \rightarrow Si_{10}^+ + Si_{10}^+ .$$

When size N increases above 30, Si_N^+ fragment both by apparent explosion down to Si_6^+-Si_{11}^+ and by the loss of one neutral atom. For example, when Si_{60}^+ is photofragmented, all the charged products are found in the range of $N = 6 \div 11$ except a small amount of Si_{59}^+. The products in the $N = 6 \div 11$ range with high intensity excitation are shown in Fig.5.6a.

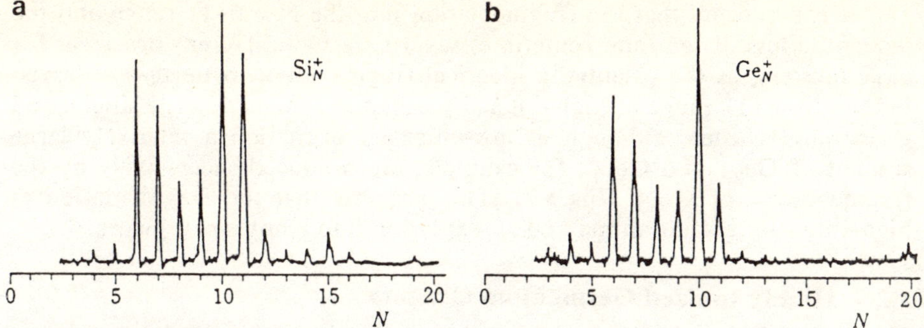

Fig.5.6. (**a**) Fragmentation products in N = 6-11 range coming from Si_{60}^+ with 43 mJ/cm² third harmonic (355nm) Nd:YAG laser; (**b**) fragmentation products in N = 6-11 range coming from Ge_{30}^+ with 10 mJ/cm² KrF (248nm) laser [5.5]

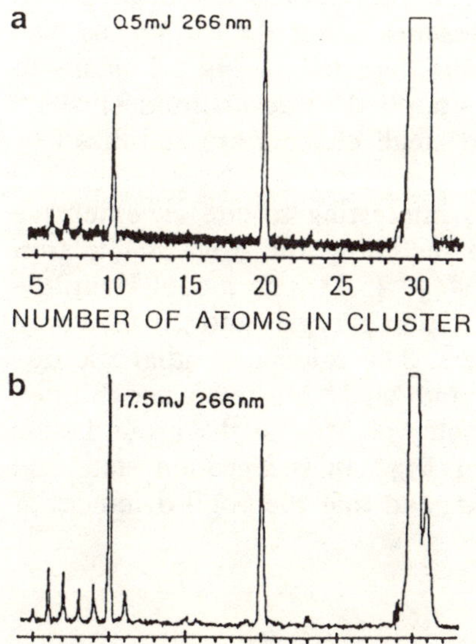

Fig.5.7. Fragmentation of Ge_{30}^+ at (a) 0.5 mJ/cm² and (b) 17.5 mJ/cm² of the fourth harmonic (266nm) Nd:YAG laser [5.5]

The smaller Ge_N^+ clusters have the same fissioning channels as Si_N^+. The larger Ge_N^+ ions exhibit the same fragmentation channels as Si_N^+: loss of single atom and fragmentation into N = 6÷11 positive cluster ions with Ge_{10}^+ dominant. As an example, the products in the N = 6÷11 range for Ge_{30}^+ with high-intensity excitation are depicted in Fig.5.6b. The intensity pattern is similar to that for Si_{60}^+ in Fig.5.6a. It is also observed that Ge_{30}^+ fragments at low-intensity excitation into almost exclusively Ge_{20}^+ presumably by loss of neutral Ge_{10}, as shown in Fig.5.7. This kind of fragmentation has not been reported for silicon.

It is suspected that the fragmentation into the N = 6÷11 range and the sequential loss of ten (and sometimes seven) are two different processes for large clusters, as the essentially identical fragmentation pattern is observed in the N = 6÷11 range at high-intensity excitation for both large silicon and germanium clusters although at low-intensity excitation a relatively large amount of Ge_{20}^+ and Ge_{10}^+, for example, are produced successively by the fragmentation of Ge_{30}^+ (Fig.5.7). This suggests that the fragmentation at high-intensity excitation may be an explosion into many small pieces.

5.2.3 Highly Ionized Germanium Clusters

The mass spectrum of germanium clusters produced by the Liquid-Metal Ion-Source (LMIS) technique was studied [5.6]. In this technique, germanium cluster ions are extracted by a strong electric field from the tip of a needle covered by molten germanium. A characteristic point of LMIS is the production of highly ionized clusters (Fig.5.8) for germanium. It is surprising to see the peaks in the mass spectrum corresponding to highly ionized clusters of small size, which are unstable if the liquid-drop model for metal clusters is assumed: fissility parameter f becomes larger than unity, as discussed in Sect.3.5.1. The experimental results depicted in Fig.5.8 seems to show that the liquid drop model cannot be applied to germanium clusters. Then a question arises how highly-ionized small clusters are stabilized or metastabilized.

In connection with this question, it is interesting to cite an extensive calculation by using a model Hamiltonian including the electron-electron interaction performed on Mg_3, Mg_3^+ and Mg_3^{++} [5.7]. The model Hamiltonian takes into account instantaneous polarization energies and the Coulomb repulsion of holes in double charged clusters. The calculated adiabatic potentials of the ground states of Mg_3, Mg_3^+ and Mg_3^{++} clusters are exhibited in Fig.5.9. The figure shows that the geometry of Mg_3 in the ground state is a regular triangle and the geometries of Mg_3^+ in the ground state and Mg_3^{++} in the metastable state are both linear, and that the bond distances of

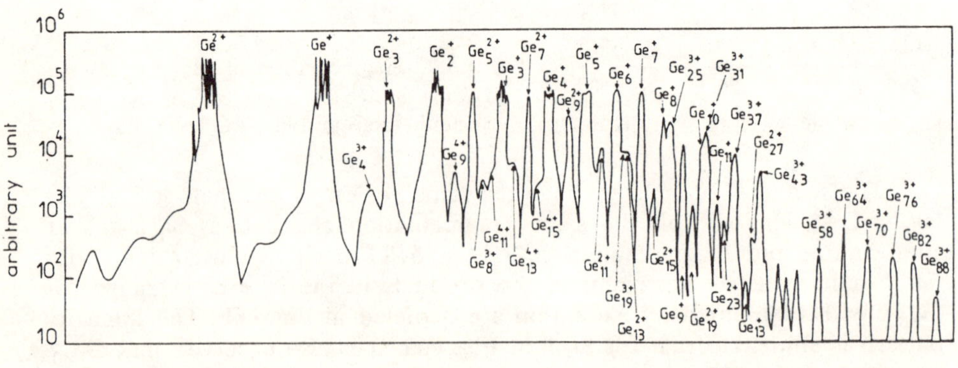

Fig.5.8. Mass spectrum of Ge_N^{p+} ions produced by the liquid-metal ion-source technique [5.6]

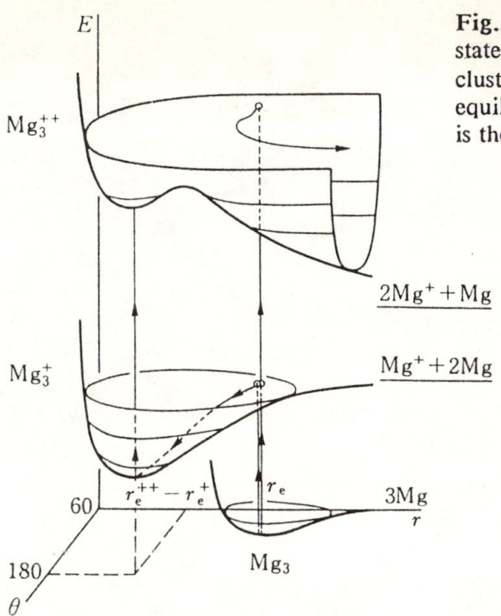

Fig.5.9. Adiabatic potentials of the ground states of Mg_3, Mg_3^+ and Mg_3^{++} clusters. The clusters are assumed to have geometries of equilateral triangles ABC: $r = r_{AB} = r_{BC}$, θ is the angle of B [5.7]

Mg_3^+ Mg_3^{++} are much shorter than that of Mg_3. This may be due to the polarization induced by positive charges localized on the central atom.

The most interesting point of the figure is that the vertical excitation of Mg_3 in the ground state to Mg_3^{++} in the electronic ground state hardly brings Mg_3^{++} into its local minimum, a metastable state, while the vertical excitation of Mg_3^+ in the ground state brings Mg_3^+ into the metastable state. Therefore, the figure seems to tell us that Mg_3^{++} in the metastable state may be produced by successive excitation $Mg_3 \rightarrow Mg_3^+ \rightarrow Mg_3^{++}$ with the relaxation of Mg_3^+.

5.3 Nonempirical Calculations for Si_6 and Si_{10}

5.3.1 Equilibrium Geometries

In Sect.5.2, Si_6 and Si_{10} are shown to be stable clusters, although no geometry has been determined and no physical property measured. The smallest six-membered ring Si_6 cluster called the chair-type cyclohexane structure (Fig.5.10a), is found in the lattice of the diamond structure. The Si_{10} cluster called the admantane structure consisting of four rings (Fig.5.11a), is also found in the diamond structure. At a glance a simple picture of the sp^3 bond seems to explain the observed stability, as the number of bonds per atom is maximum at these structures [5.9]. However, in what follows, we shall show that the sp^3 bond picture is too simple to explain a large reconstruction theoretically derived: the reconstruction is caused by the interaction among many dangling bonds inherent in microclusters.

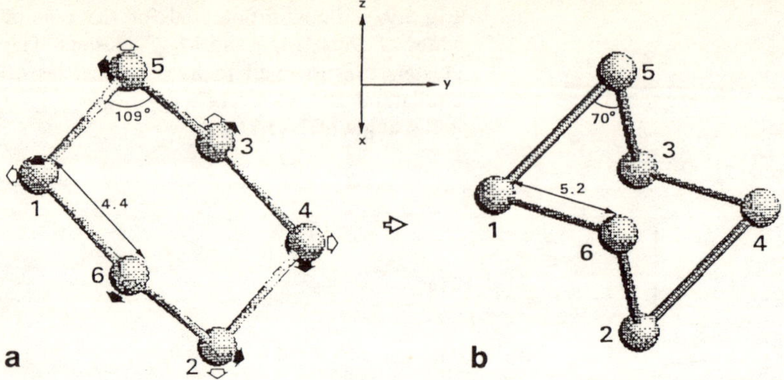

Fig.5.10. (a) The chair-type cyclohexane structure originally assumed for Si_6; (b) The reconstructed structure close to an octahedron [5.8]

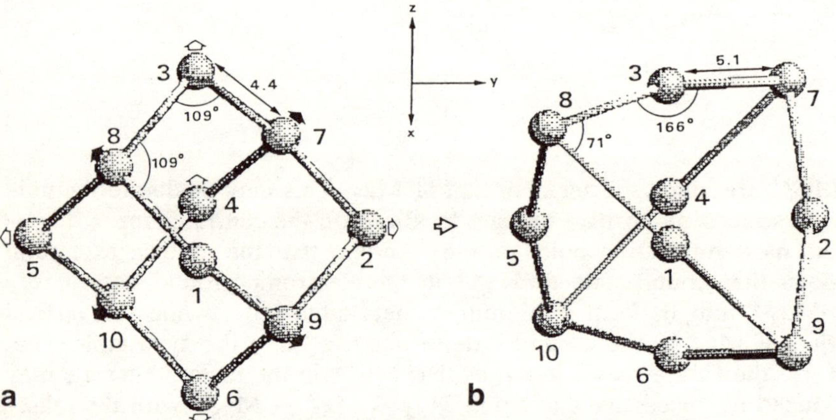

Fig.5.11. (a) The admantane structure originally assumed for Si_{10}; (b) The reconstructed structure close to a tetrahedron [5.8]

The non-empirical calculation of the stable reconstructed structure of Si_6 and Si_{10} cited here [5.8] is to calculate forces acting on the constituent atoms by using the local-density-functional scheme with the Linear-Combination of Atomic-Orbitals $X\alpha$-force method, i.e. by using (4.3). This method is simply called the LCAO-$X\alpha$-force method. In the calculation, the structures assumed at the beginning are those found in the diamond structure with the bond length equal to 4.4 a.u. (Figs.5.10a and 11a). The equilibrium geometry is determined from the condition that the force must vanish. It is assumed that each cluster retains its original symmetry, D_{3d} for Si_6 and T_d for Si_{10}. As far as we are concerned with physical or qualitative arguments on the reconstruction, this assumption seems to be justified, as the coupling of valence electrons with a symmetry-conserving displacement mode (a breathing mode) is normally the strongest. Actually a qualitatively similar result has been obtained in a different type of non-empirical calcu-

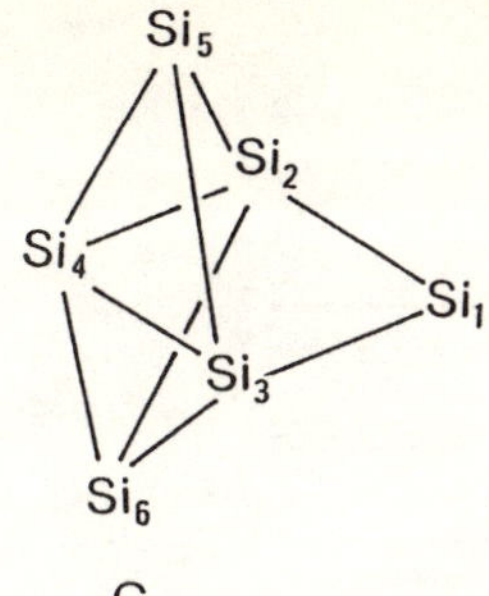

Fig.5.12. The ground state geometry of Si_6 calculated without symmetry restriction [5.10]

lations without symmetry restriction, as depicted in Fig.5.12 for Si_6. The structure in the figure can be derived from a tetragonal bipyramid close to an octahedron by distorting the apex atoms along a direction parallel to the diagonal of the square. The same calculation without symmetry restriction for Si_{10} predicts a tetra-capped octahedral structure of the T_d symmetry which is close to that given in Fig.5.11b [5.10].

The symmetry-restricted non-empirical calculation by the use of the LCAO-Xα-force method gives the equilibrium geometry of Si_6 close to an octahedron, as shown in Fig.5.10b in qualitative agreement with the geometry obtained by the symmetry-unrestricted calculation. The same symmetry-restricted calculation for Si_{10} gives an equilibrium geometry close to a tetrahedron containing an octahedron almost in its inside (Fig.5.11b). This structure may be called a tetra-capped octahedron in qualitative agreement with the geometry obtained by the symmetry-unrestricted calculation as already mentioned. It should be noticed that the bond length is elongated up to 5.2 a.u. and the regular triangles called the (111) triangles, (1,2,3) and (4,5,6), contained in Si_6 and Si_{10}, as seen in Figs.5.10 and 11, are contracted by 20% in length after the reconstruction. In what follows, we shall argue that this contraction is due to the interaction among the dangling bonds perpendicular to the triangle. Anyhow, the elevation of approximate symmetry, a chair-type cyclohexane → an octahedron for Si_6 and an admantane → a tetrahedron, found in the non-empirical calculation is quite interesting and seems to suggest the existence of a simple model for discussing the bonding of semiconductor clusters.

5.3.2 Energy Levels

It was pointed out in the previous subsection that the quilibrium geometry of Si_6 is close to an octahedron. In Fig.5.13, the calculated one-electron energy levels of the reconstructed Si_6 (Fig.5.10b) and an octahedral Si_6 with bond length 5.44 a.u. are compared. In the octahedral Si_6, the highest occupied molecular orbital T_{1u} is partially filled. Therefore, the reconstructed structure of Si_6 may be regarded as an octahedron distorted by the Jahn-Teller effect. In the reconstructed Si_6, the energy gap between the lowest unoccupied A_{2u} and the highest occupied E_u levels is still small. This suggests that a further Jahn-Teller distortion for nearly degenerate levels may

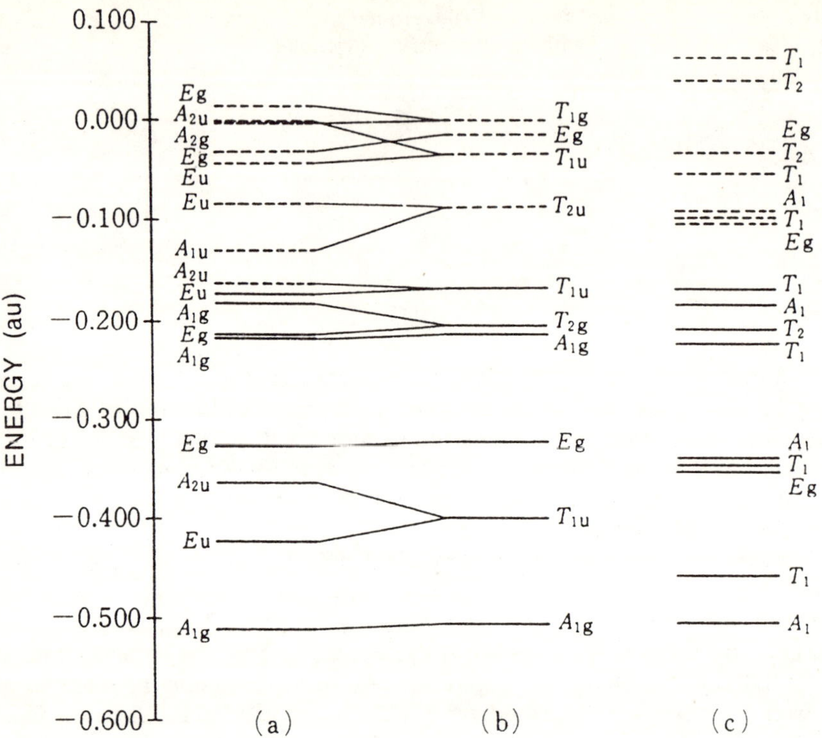

Fig.5.13. Electronic energy levels of **(a)** the reconstructed Si_6, **(b)** the octahedral Si_6 with bond length of 5.44 a.u., and **(c)** the reconstructed Si_{10}. Solid lines represent the occupied levels, while broken lines the unoccupied [5.8]

be expected. This is realized in the result of the symmetry-unrestricted calculation for Si_6 (Fig.5.12) where the symmetry of the equilibrium structure is C_{2v}, while it is D_{3d} in the result of symmetry-restricted calculation. Besides the Jahn-Teller distortion, the small energy gap between the lowest unoccupied and the highest occupied levels raises a question whether or not the Si_6 cluster could be called a semiconductor.

The calculated energy levels of the reconstructed Si_{10} are also exhibited in Fig.5.13. The highest occupied T_1 level is completely filled, and a large energy gap is seen between the lowest unoccupied and the highest occupied levels. Actually, no further distortion is found in the symmetry unrestricted calculation as already mentioned.

5.3.3 Density Distribution of Electrons

In order to understand the nature of chemical bonds in the reconstructed clusters, it is important to examine electron densities in the Highest Occupied Molecular Orbital (HOMO). These electron densities in the starting and the reconstructed structures of Si_6 and Si_{10} in the plane of the (1,2,3) triangle are depicted in Fig.5.14. It is seen in the figure that the electron den-

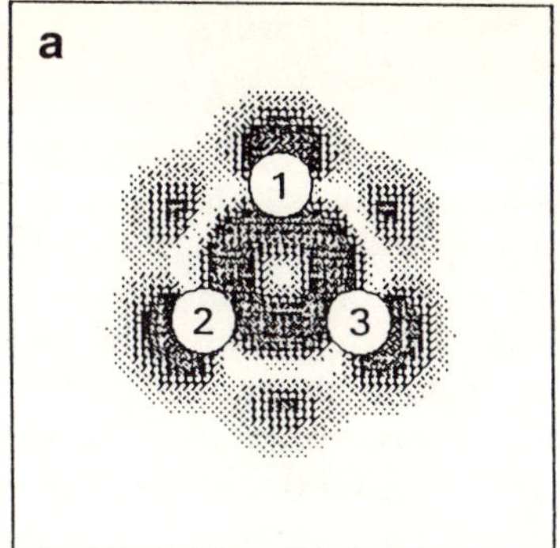

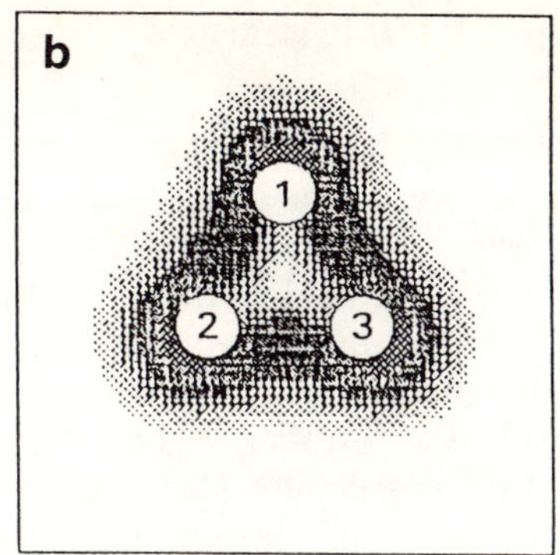

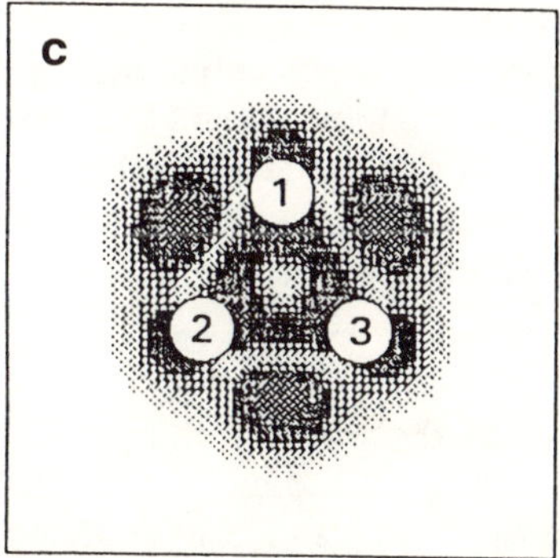

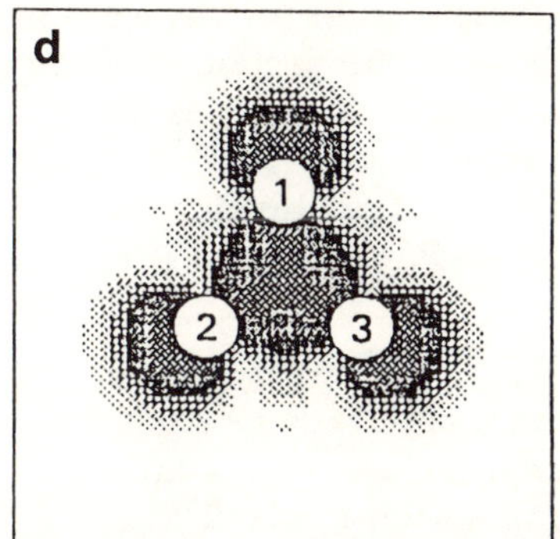

Fig.5.14. Electron densities of the highest occupied molecular orbitals in the (1,2,3) triangle plane for (**a**) the starting structure of Si_6, (**b**) the reconstructed Si_6, (**c**) the starting Si_{10}, and (**d**) the reconstructed Si_{10}. The darker the map, the higher the density [5.8]

sities of both the Si_6 and Si_{10} clusters, which resemble each other in the starting structures, change into more strongly connected ones in the reconstructed structures. In the reconstructed structures, atoms in the triangle are connected along the triangle edges in Si_6 and through the triangle center in Si_{10}.

These peculiar electron densities in the reconstructed clusters and the reconstruction into the structure of higher symmetry seem to suggest the presence of a new type of chemical bond in semiconductor clusters which is different from the covalent bond in the group IV semiconductor crystals. Unfortunately, at present, we have no simple picture of chemical bonds in the semiconductor clusters. Peculiar properties of semiconductor clusters concerning chemical bond are further discussed in the next section by citing a non-empirical calculation for six-membered-ring silicon clusters of larger size.

5.4 Force and Virial Analysis of Larger Silicon Clusters

5.4.1 Six-Membered-Ring Clusters

In what follows, we call a chair-type cyclohexane (Fig.5.10a) a six-membered-ring cluster. When a pyramidal cluster consisting of four atoms sits on this ring, being bonded with three dangling bonds of one of the (111) triangles, as seen in the derivation of Si_{10} from Si_6, the number of six-membered rings increases by three. This large increase of the number of the rings means that the number of bonds per atom is maximum again at the newly derived structure. This fact is pointed out to discuss that, if Si_N have the geometries found in the diamond structure, they are expected to show magic numbers with the period, four, like 6, 10, 14, 18, 22, 26 etc. The structures of these clusters are depicted in Fig.5.15.

The purpose of this section is to demonstrate that the peculiar properties found for Si_6 and Si_{10}, for example a large contraction of the (111) triangle, as displayed in Sect.5.3.1, are rather general for the six-membered-ring silicon clusters. For this purpose, the non-empirical calculation by use of the LCAO-Xα-Force method is extended to larger clusters of this series.

5.4.2 Triangle Contraction

In order to examine whether or not the concept of the triangle contraction may be generalized, the forces are calculated acting on atoms in the six-membered-ring silicon clusters, Si_6, Si_{10}, Si_{14}, Si_{18}, Si_{22} and Si_{26} as taken out from the diamond structure. In this calculation atomic orbitals up to the 3d are used for the basis functions. All the bond lengths are assumed to be the same as that of the bulk, 4.44 a.u. Symmetries of Si_{14}, Si_{18}, Si_{22} and Si_{26} are D_{3d}, C_{2v}, C_{3v} and T_d, respectively.

Using the calculated forces, we evaluate the virial, which is related to a tendency for the cluster to expand or contract. The virial for a cluster, V, is given by

$$V = \sum_i \mathbf{f}_i \cdot \mathbf{r}_i \,, \tag{5.1}$$

where $\mathbf{f}_i$ is the force acting on the ith atom at position $\mathbf{r}_i$. The virial V is independent of the position of the origin of cartesian coordinates, as

$$\sum_i \mathbf{f}_i = 0 \,. \tag{5.2}$$

If V is positive (negative), the cluster tends to expand (constract). Virials for Si_N ($N = 6,10,14,18,22,26$) have been calculated as given in Table 5.3. All the calculated virials are negative, showing that the six-membered-ring silicon cluster with the crystalline bond length tend to contract. Table 5.3 also reveals that the virial per atom, V/N, takes a maximum value at $N = 14$

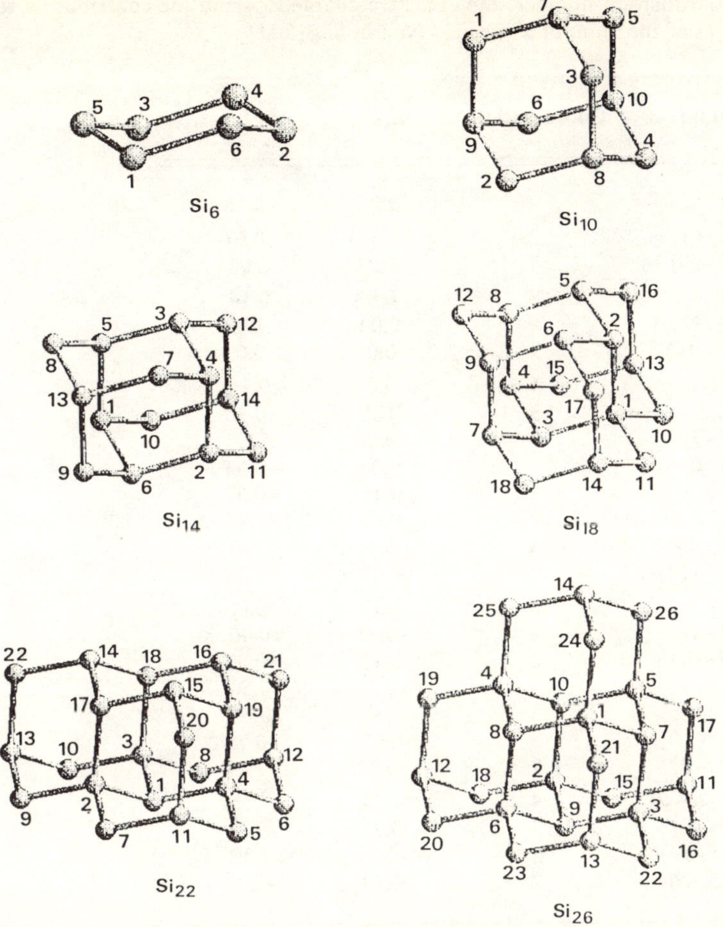

Fig.5.15. The structure of six-membered-ring clusters as taken from the diamond structure [5.11]

Table 5.3. Virials and virials per atom of Si_N in a. u. [5.11]

N	V	V/N
6	-1.08	-0.18
10	-3.70	-0.37
14	-8.22	-0.59
18	-8.62	-0.48
22	-10.01	-0.46
26	-3.36	-0.13

Table 5.4. The coordination number, the effective charge Z_{eff} and the contribution to the virial $f_i \cdot (r_i - r_G)$ of the atom of a specific type in Si_N [5.11]

N	Type of atom		Coordination	Z_{eff}	$f_i \cdot (r_i - r_G)$
6	I	(1-6)[a]	2	0.0	-0.18
10	I	(1-6)	2	0.07	-0.67
	II	(7-10)	3	-0.11	0.08
14	I	(1-6)	3	-0.03	0.14
	II	(7-12)	2	0.04	-1.00
	III	(13,14)	3	-0.01	-0.69
18	I	(1)	4	-0.04	0.44
	II	(2,3)	3	0.08	-0.23
	III	(4-7)	3	-0.08	-0.07
	IV	(8,9)	3	0.04	-0.53
	V	(10,11)	2	0.11	-0.57
	VI	(12)	2	0.03	-0.86
	VII	(13,14)	3	-0.14	0.13
	VIII	(15-18)	2	0.08	-1.38
22	I	(1)	3	0.13	-0.05
	II	(2-4)	4	-0.21	0.06
	III	(5-10)	2	0.11	-0.96
	IV	(11-13)	3	-0.18	0.50
	V	(14-16)	3	-016	0.63
	VI	(17-19)	3	0.17	-0.78
	VII	(20-22)	2	-0.12	-1.81
26	I	(1-6)	4	-0.22	1.18
	II	(7-10)	3	0.27	-0.31
	III	(11-14)	3	-0.23	1.39
	IV	(15-26)	2	0.14	-1.23

[a] The number in parentheses are the atom numbers indicated in Fig.5.15

and decreases for larger clusters. A fairly small value of V/N for Si_{26}, suggests that a tendency to take the bulk geometry rather than a close-packed geometry for small clusters appears around the size N = 26.

In Table 5.4, we give the effective charge of Z_{eff} of atoms, i.e.,

$$Z_{eff} = 14 - z_i , \tag{5.3}$$

where z_i is the Mulliken charge of the ith atom calculated from self-consistent electron densities, and the calculated contributions to the virials, $f_i \cdot (r_i - r_G)$, for each type of atoms in Si_N together with their coordination numbers. Here, r_G represents the position of the center of gravity of each cluster. The contribution to the virial indicates that the ith atom has a tendency of going outwards (inwards) if it is positive (negative). In the table one can see a strong correlation between the effective charge and the contribution to the virial with a few exceptions: the constituent atom is pushed outwards (inwards) if its effective charge is negative (positive). This correlation may be understood by considering bond angles, 120°, 109° and 90°, of the sp^2, sp^3 and s^2p^3 hybridizations, respectively. The silicon atom having excess electrons will have the s^2p^3 component in addition to the main sp^3 component that pushed the atom outwards by making the bond angles smaller than 109°. On the other hand, the positively charged silicon atom will have some sp^2 component that pushes the atom inwards by increasing the bond angle towards 120°.

In order to clarify the existence of the triangle contraction, we calculate V_t for the (111) triangles defined as

$$V_t = \sum_i{}' \mathbf{f}_i \cdot (\mathbf{r}_i - \mathbf{r}_c) , \tag{5.4}$$

where the sum is over the three atoms in a (111) triangle, and $\mathbf{r}_c$ is the position vector of the center of the triangle. In the same way as for the virial of a whole cluster, the positive (negative) V_t means that the triangle considered tends to expand (contract).

The calculated V_t of Si_N are given in Table 5.5. In Si_6, Si_{10}, Si_{14} and Si_{26}, only one type, A, of the (111) triangles exists. In Si_{18}, however, there are three types of the triangles, A, B and C. Table 5.5 reveals that, except for the A type of Si_{22}, all the triangles tend to contract. The exceptional expansion of the A type triangle may be caused by the contraction of the B and C type triangles of Si_{22}.

Table 5.5. The calculated V_t for the (111) triangles of Si_N in atomic units. Different types of the triangles are indicated as A, B, and C. The triangle is specified by three constituent atoms whose numbers (Fig.5.15) are given in parentheses [5.11]

N	Types of triangles	Number of triangles	V_t
6	A (1,2,3)	2	-1.33
10	A (1,2,3)	4	-1.33
14	A (4,11,12)	6	-1.16
18	A (4,7,12)	2	-0.41
	B (8,15,16)	2	-1.23
	C (3,10,15)	4	-0.89
22	A (14,15,16)	1	0.27
	B (7,17,20)	6	-1.11
	C (1,5,7)	3	-0.86
26	A (7,16,17)	12	-0.96

6. Rare-Gas Clusters

Rare-gas clusters are ideal microclusters, for which a reliable theoretical treatment may be achieved due to the applicability of a simple pair potential such as the Lennard-Jones potential. Dynamical and thermodynamical properties of rare-gas clusters have already been discussed in Sects.2.2 and 2.4.1. In this chapter we first treat static properties of xenon and argon clusters observed by mass analysis. Then, we discuss the ground states of ^{4}He and ^{3}He clusters which are rather peculiar among rare-gas clusters.

6.1 The Magic Numbers for Packing

6.1.1 Xenon Clusters

Xenon clusters are nucleated in a supersaturated phase, a supersonic jet, obtained by adiabatic expansion of xenon through a capillary into a vacuum. The jet is skimmed by two conical collimators and enters the time-of-flight mass spectrometer, where it is ionized by a pulsed electron beam with an energy of 30 eV [6.1].

Figure 6.1 shows the mass spectrum thus observed, which is subdivided into two overlapping ranges. In the figure, we observe local maxima and the sudden decrease thereafter in the intensity. The most prominent local maxima appear at the sizes N = 13, 19, 25, 55, 71, 87 and 147. These numbers can be called *magic numbers*. The less prominent magic numbers are found for N = 23, 81, 101 and 135, as indicated in brackets.

The magic numbers 13, 55 and 147 may be identified to be the so-called *icosahedral numbers* I_K for integers K given as [6.2]

$$I_K = \frac{1}{3}(10K^3 - 15K^2 + 11K - 3) \,, \tag{6.1}$$

which are the numbers of atoms forming an icosahedron with K atomic shells. However, the other prominent magic numbers 19, 25, 71 and 87 remain unidentified.

6.1.2 Argon Clusters

Consider an experiment, where positive ions are first produced upstream from the nozzle by a corona discharge and swept into the jet. There they become condensation nuclei and charged clusters grow around them [6.3].

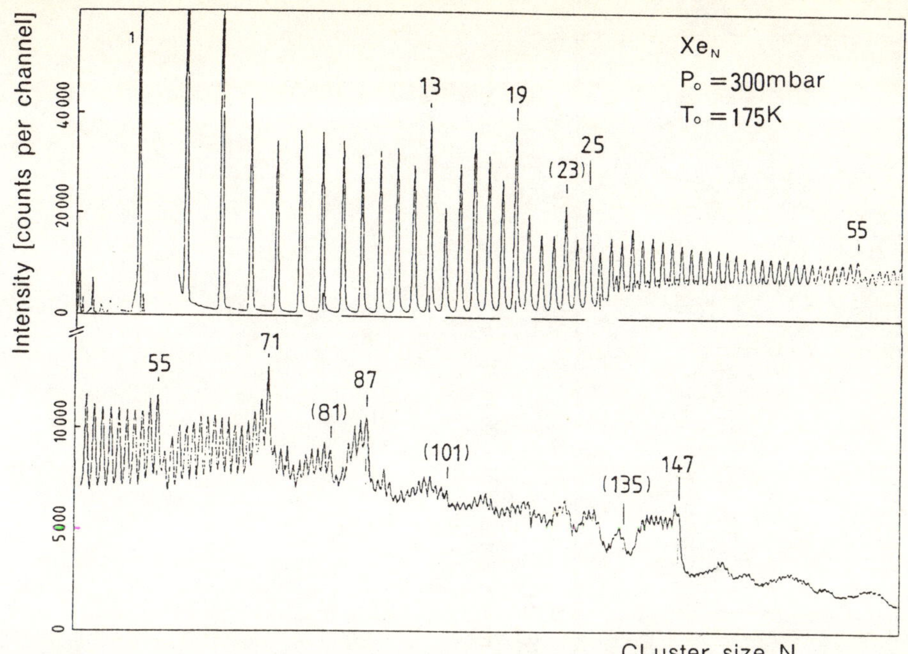

Fig.6.1. The observed mass spectrum of xenon clusters. The magic numbers are indicated in boldface. Numbers in brackets are the less pronounced magic numbers [6.1]

In this case, the formation of clusters is described in terms of a competition between the process of heating due to accretion and that of cooling due to evaporation. The cluster ions thus produced are skimmed from the central portion of the jet, accelerated, and passed through additional stages of differential pumping into a magnetic mass spectrometer. It is important for the clusters to drift for a relatively long time before entering the spectrometer.

The mass spectrum observed is displayed in Fig.6.2, ranging from the sizes N = 10 to N = 90. The magic numbers of this spectrum, 13, 19, 23, 55, 71, 81 and 87, agree with those of the Xenon clusters exhibited in Fig.6.1, although the spectrum of Fig.6.2 is richer in structure.

To understand the experimental mass spectrum, the following semi-quantitative model is employed [6.3]. It assumes that the structure in the mass distribution is formed in the evaporation process after accretion in the jet ceases. Let us describe the distribution of hot clusters by $n(\overline{N},E)$, the number of $\overline{N}$-atom clusters per unit volume with total energy in the range from E to E+dE. We further assume that in a time short in comparison to the $2\cdot10^{-4}$ s drift time the clusters evaporate and cool until they reach a bound ground state. If we neglect the residual kinetic energy of the evaporated atoms, every cluster with $-E_b(N) \leq E \leq -E_b(N-1)$, $E_b(N)$ being the binding enery of an N-atom cluster, ends up as a bound N-atom cluster. Thus the final number of N-atom clusters per unit volume, $I_0(N)$, is given as

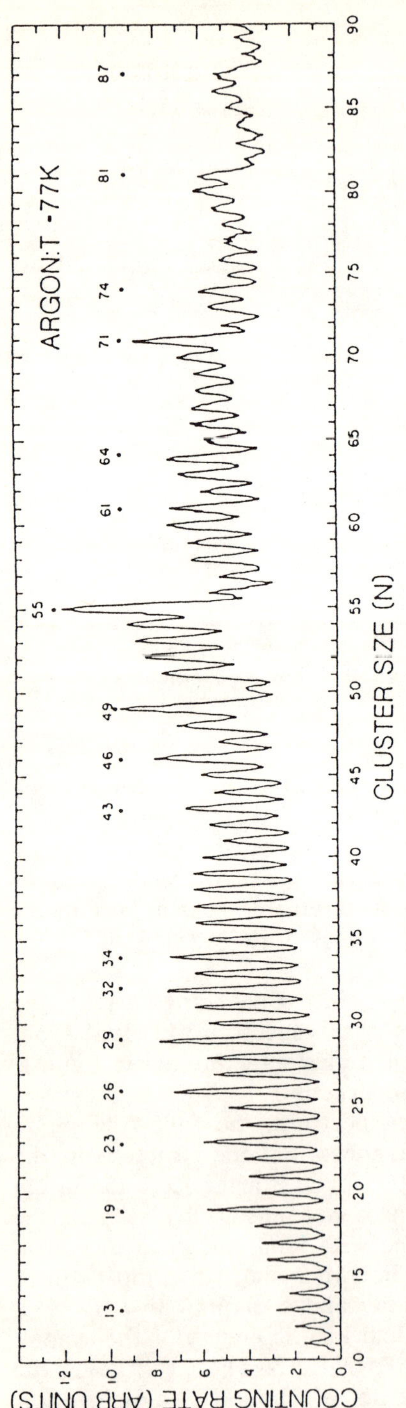

Fig.6.2. The observed mass spectrum of charged argon clusters ranging from N = 10 to N = 90 [6.3]

$$I_0(N) = \sum_{\bar{N}=N}^{\infty} \int_{-E_b(N)}^{-E_b(N-1)} n(\bar{N},E)dE \,. \tag{6.2}$$

If $n(\bar{N},E)$ is assumed to be a slowly varying function of E, factorizing $n(\bar{N},E)$ outside the integral one obtains

$$I_0(N) = \Delta E_b(N) \sum_{\bar{N}=N}^{\infty} n(\bar{N},E_b(N)) \,, \tag{6.3}$$

where $\Delta E_b(N) = E_b(N)-E_b(N-1)$. The sum in (6.3) is a slowly varying function of N determined by conditions in the jet. Equation (6.3) shows that the fine structure of the mass spectrum is nearly proportional to the successive binding-energy differences. This should be compared with the argument that one of the measures for the relative stability of clusters yielding the fine structure of the mass spectrum is given by the second derivative of the total energy with respect to N, as shown in (3.49). Both the first and the second derivatives of the total energy can determine the magic numbers although the second derivative enhances the fine structure.

It seems to be clear that the magic numbers 13 and 55 correspond to the closing numbers of the first and the second shells of the icosahedral packing, respectively. To understand the magic numbers in the range of N = 13 to 55, we start with the assumption that the core structure is a rigid 13-atom cluster with icosahedral symmetry and the second shell is formed by decorating this core with atoms at sites of high symmetry until a 55-atom icosahedron is completed. We assume that these sites of high symmetry are those above the centers of Faces of the Core (FC sites) with three near neighbors in the core, those above the middle of the Edges of the Core (EC sites) with two, and those above Vertices of the core (V sites) with one. Occupation of an FC site precludes occupation of adjacent EC sites, and vice versa; but occupation of a V site is compatible with either one. Consequently there are two mutually exclusive lattices supported by the core: the FCV lattice consisting of 20 FC and 12 V sites, and the ECV lattice (the second shell of an icosahedron) consisting of 30 EC and 12 V sites. These lattices are schematically illustrated in Fig.6.3, where V sites are indicated by thick circles: the outer-most ring is a V site.

The ECV lattice are favored when the second shell is nearly complete because it contains more sites, but the FCV will be favorable when the shell is fairly empty because it optimizes the intershell interaction. Our problem is to find the distribution of a given number of atoms on each of these lattices with the minimum energy. To do this, we adopt a drastic assumption for the atomic interactions. Namely, we assume that both the intershell and intrashell interactions act only on the nearest-neighbor intershell and intrashell pairs, respectively, and their strengths are the same for each lattice.

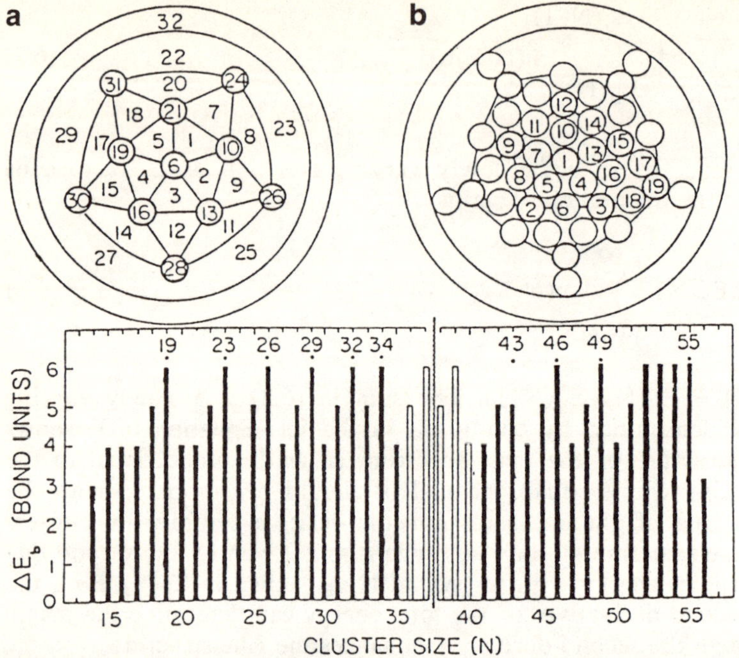

Fig.6.3. Schematic figures and calculated binding-energy differences for (a) FCV and (b) ECV lattices. Heavy circles represent sites above vertices of the underlying icosahedral core [6.3]

Validity of the assumption will be confirmed later by showing its capability of explaining the observed structure of the mass spectrum.

Our assumption for the atomic interactions makes $\Delta E_b(N)$ be proportional to the amount of increase in the total number of the nearest-neighbor intershell and intrashell pairs when the Nth atom is added. It is rather evident that, when the second shell is fairly empty, the distribution optimizing the interaction energy is obtained by placing atoms on the FCV lattice in the sequence of numbers indicated in Fig.6.3a. The figure tells us that, when atom 1 is placed at the FC site, its nearest-neighbors are three atoms of the core cluster below the V sites, 6, 10 and 21. When atom 2 is placed in addition to atom 1, its nearest-neighbors are three core atoms below 6, 10 and 13 plus atom 1 already placed, giving the increase in the total number of nearest-neighbor pairs by four. The increment of the total number of the pairs is shown in the bar graph below the FCV lattice in Fig.6.3a.

On the other hand, when the second shell is nearly complete, the distribution of atoms optimizing the interaction energy is obtained by taking away atoms from the completely filled ECV lattice in the sequence of numbers indicated in Fig.6.3b. The figure tells us that, when atom 1 is taken away from the complete ECV lattice, one bond with a core atom below the V site 1 plus five bonds with atoms at the EC sites, 4, 5, 7, 10 and 13 are broken. When atom 5 is taken away in addition to atoms 1, 2, 3 and

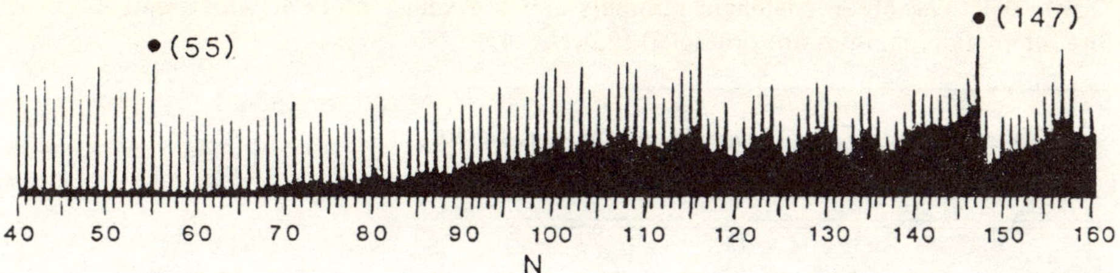

Fig.6.4. The experimental structure function for $40 < N < 160$. The intensity is given in arbitrary units [6.4]

4, two bonds with two core atoms below sites 1 and 2 plus three bonds with atoms 6, 7 and 8 are broken, giving the decrease in the total number of the pairs by five. The decrement of the total number of the pairs when the cluster size N decreases by one is illustrated by the bar graph below the ECV lattice figure in Fig.6.3b.

Since the final number of N-atom clusters per unit volume, $I_0(N)$, is proportional to $\Delta E_b(N)$ according to our model for the formation of bound clusters, anomalies in the bar graph in Fig.6.3 should be compared with those observed in the mass spectrum in Fig.6.2. The correspondence is complete, if we adopt the FCV lattice for $13 < N < 35$ and the ECV for $40 < N \leq 55$. This seems to confirm an approximate validity of our assumption on the atomic interactions in the model for discussing relative abundance of clusters in a limited range of size, although our assumption is too crude to tell in what range the FCV or ECV lattice distribution can be adopted.

Figure 6.4 depicts the experimental structure function representing in some way the structure of the mass spectrum in the size range from $N = 40$ to 160 [6.4]. The principal features of the structure function are the anomalies, large values of $I(N)/I(N+1)$, at certain values of N (magic numbers), which are listed in Table 6.1 for the range of $55 < N \leq 147$. The magic number $N = 147$ is clearly the closing number of the third shell of an icosahedron. To explain the observed magic numbers in the range of $55 < N < 147$, we use the same model adopted for explaining those in the range of $13 < N < 55$. The FCV and ECV lattices in the third shell, denoted as FCV3 and ECV3, are exhibited in Fig.6.5. The FCV3 lattice consists of $3\times 20 = 60$ FC sites and 12 V sites, and the ECV3 $(2\times 30)+20 = 80$ EC sites and 12 V sites giving 92 sites of the third shell of an icosahedron.

In just the same way, as done for filling the second shell, the distribution optimizing the interaction energy in the third shell is obtained by placing atoms on the FCV3 lattice sites when the third shell is fairly empty. The anomalies of the interaction energies may at least be expected at the following values of N:

i) $N = 55+16 = 71$, when 15 FC sites and a central V site of a pentagonal pyramid are filled.

ii) $N = 71+10 = 81$, when 9 FC sites are empty, and a central V site of the second pentagonal pyramid topped at one of the five corners of the first pentagonal pyramid is filled.

Table 6.1. The observed magic numbers and the values of N at which anomalies of the interaction energies are predicted ($55 \leq N \leq 147$)

Observed		Magic number predicted	
edge	minima[a]	FCV3	ECV3
55		55	
61[b]			
64[b]			
71		71	
74[b]			
81		81	
87		88	
	102		
	105		
	112		
116			116
	120		119
	126		125
	132		131
	137		137
147			147

[a] The magic numbers may be those given by these numbers minus one
[b] Less obvious but reproducible edges

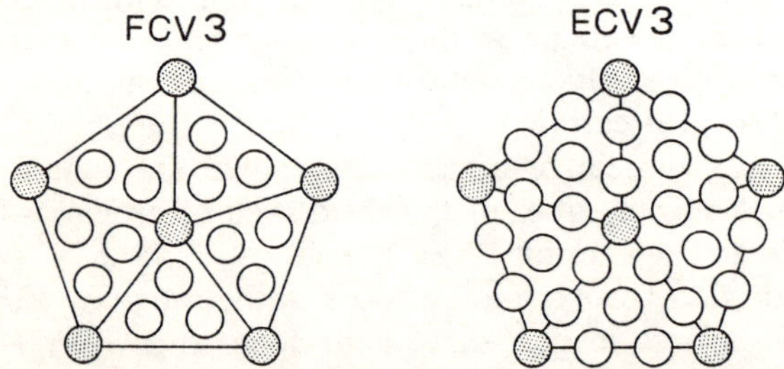

Fig.6.5. Schematic figures of the FCV and ECV lattices in the third shell [6.4]

iii) $N = 81+7 = 88$, when 6 FC site are empty, and a central V site of the third pentagonal pyramid topped at one of the two common corners of the first and the second pentagons are filled.

These values of N are compared with the observed magic numbers in Table 6.1.

When the third shell is nearly complete, the distribution optimizing the interaction energy is obtained by taking away atoms from the completely filled ECV3 lattice. The anomalies of the interaction energies may, at least, be expected at the following values of N:

i) N = 147-10 = 137, when 10 atoms on one of the five triangle faces of a pentagonal pyramid are taken away.
ii) N = 147-16 = 131, when 16 atoms on the neighboring two of the five triangle faces are taken away.
iii) N = 147-22 = 125, when 22 atoms on the neighboring three faces are taken away.
iv) N = 147-28 = 119, when 28 atoms on the four faces of a pentagonal pyramid are taken away.
v) N = 147-31 = 116, when all the atoms on a pentagonal pyramid are taken away.

These values of N are compared with the observed magic numbes in Table 6.1.

Comparisons of the magic numbes predicted by the present model with those observed up to N = 147 seem to offer strong support for an icosahedrally derived shell structure for charged argon clusters. Furthermore, low-resolution mass spectra have been observed for the clusters up to N = 561, the closing number of the fifth shell, it has also been discussed that the present model seems applicable for clusters up to N = 309, the closing number of the fourth shell [6.4].

6.2 Helium Clusters

6.2.1 Magic Numbers

Clusters of ^{4}He and ^{3}He atoms formed in a supersonic free jet expansion are studied by electron-bombardment ionization and mass spectrometry [6.5]. The observed mass spectra of these clusters are displayed in Fig.6.6. In the figure, magic numbers are found at N = 7, 10, 14, 23 and 30 for ^4He_N and at N = 7, 10, 14, 21 and 30 for ^3He_N. It is uncertain whether the mass spectra reflect the size distribution of neutral clusters or the distribution determined by fragmentation in the ionization processes.

The size distribution of ^{4}He cluster ions produced without ionization processes is studied by using an injected-ion drift tube mass spectrometer [6.6]. A drift tube is set in a cryogenic system and cooled with liquid helium. The injected helium ions He^+ are quickly thermalized by elastic collisions with helium atoms as drifted toward the end of the tube by a uniform electric field E applied in the tube. The ions ejected from the exit aperture at the end of the tube are focussed and mass analyzed with a quadrupole mass spectrometer.

The mean energy of drifting ions can be controlled by changing the electric field strength E and the number density of the gas ρ. The effective temperature of the colliding system is defined as

$$T_{eff} = \frac{2}{3k_B}\langle E_r\rangle \ , \tag{6.4}$$

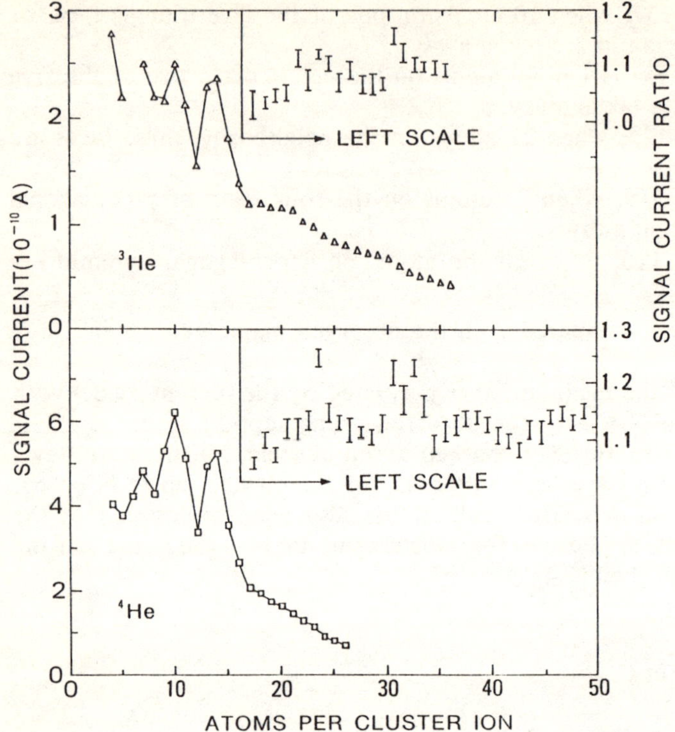

Fig.6.6. Electron multiplier output currents from cluster ions of ^{3}He at temperature T_0 = 3.2 K, stagnation pressure P_0 = 550 Torr, and ^{4}He at T_0 = 4.2 K, P_0 = 400 Torr. Inserts show ratios of adjacent peak amplitudes [6.5]

where $\langle E_r \rangle$ is the mean relative energy given as

$$\langle E_r \rangle = \frac{1}{2} M V_d^2 + \frac{3}{2} k_B T_g \,. \tag{6.5}$$

In (6.5), M is the mass of a He atom, V_d the drift velocity of a He_N^+ cluster ion, and T_g the temperature of the gas when E = 0. Then, the effective temperature is given by

$$T_{eff} = T_g + \frac{M}{3k_B} \left(\frac{K_0 E}{\rho} \right)^2 , \tag{6.6}$$

where the zero-field-reduced mobility K_0 is used instead of V_d: K_0 can be roughly estimated by using the Langevin theory. When T_{eff} is increased by increasing E, the clusters with $\Delta E_b(N) = E_b(N) - E_b(N-1)$ smaller than $k_B T_{eff}$ become unobservable because of the dissociation of these clusters by collisions; here $E_b(N)$ is the binding energy of an N-atom cluster. Therefore, the drift field strength, which makes the ion signal of an N-atom cluster disappear, is called the disappearance field strength for the N-atom cluster. It is evident that the disappearance field strength reflects the stability of a cluster ion.

The disappearance field strengths experimentally observed for He_N^+ at three values of the gas pressure are shown in Fig.6.7. The disappearance field for He_3^+ is too high to be measured. This fact tells us that the trimer is very stable as compared with the larger cluster showing a drastic change in the binding properties between the clusters of N = 3 and 4. In Fig.6.7, stepwise changes of the disappearance field are observed between N = 10 and 11 for all the gas pressure and between N = 14 and 15 for the gas pressure of 0.06 Torr. The results indicate the fact that the binding energy of He_N^+ abruptly decreases at N = 11 and 15. Therefore, N = 10 and 14 may be the magic numbers of He_N^+. These magic numbers coincide with those observed for neutral helium clusters in a supersonic nozzle beam [6.5]. This suggests that the magic numbers for the neutral clusters reflect the size distribution realized after fragmentation in the ionization space of the mass spectrometer.

No direct observation of the structures of stable He_{10}^+ and He_{14}^+ cluster ions has been made yet. It is, however, speculated that a He_2^+ cluster sits at the center of these stable cluster ions as a tightly bound molecular ion and

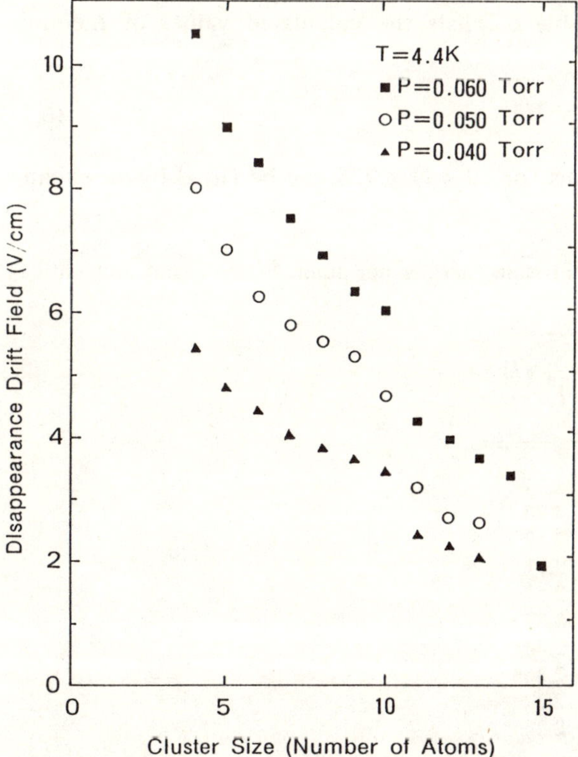

Fig.6.7. Disappearance drift field strength where the ion signal of He_N^+ disappears when the drift field strength is increased by keeping the gas pressure constant [6.6]

the remaining 8 and 12 He atoms for He_{10}^+ and He_{14}^+ are weakly bound by the polarization force forming the closed first shells of a cube and an icosahedron, respectively.

6.2.2 Nonempirical Calculation

Quantum mechanical calculations of the ground state of a ^4He_N cluster are made by using the Green's Function Monte-Carlo method (GFMC) for $3 \leq N \leq 112$ and the Variational Monte-Carlo method (VMC) for $4 \leq N \leq 728$ [6.7]. The GFMC calculation may, in principle, be considered to be exact aside from the statistical errors which may be made arbitrarily small [6.8]. The VMC calculation uses a variational wave function which can take into account two-body and three-body correlations. In both calculations a Hartree-Fock interatomic potential is used.

The ground-state energies per atom, E(N)/(N), obtained by the GFMC and VMC calculations are listed in Table 6.2. The statistical errors in the VMC are comparable to those of the GFMC results. The VMC calculations give an upper bound to the ground-state energy, while the GFMC provide exact results. The closeness of the VMC and GFMC results indicates good quality of the variational wavefunction. The three-body correlations are found to contribute about 13% to E(N)/N for the larger droplets.

The last column of Table 6.2 lists the calculated values of the unit radii, $r_0(N)$, defined by

$$r_0(N) = (5/3)^{1/2} \langle r^2 \rangle^{1/2} N^{-1/3} . \qquad (6.7)$$

The VMC $r_0(N)$ in Ångstroms for $20 \leq N \leq 728$ can be fitted by an expansion in powers of $N^{-1/3}$ as

Table 6.2. The calculated ground-state energies per atom, E(N)/N, and unit radii r_0 for ^4He_N [6.7]

N	E(N)/N [K] GFMC	E(N)/N [K] VMC	r_0 [Å] GFMC
3	-0.039		5.35
4	-0.133	-0.128	4.20
8	-0.616	-0.597	3.19
20	-1.627	-1.570	2.71
40	-2.49	-2.396	2.57
70	-3.12	-3.02	2.47
112	-3.60	-3.52	2.44
240		-4.19	2.36[a]
728		-4.95	2.32[a]
∞	-7.11	-6.88	2.22

[a] VMC values

$$r_0(N) = 2.24 + 0.38\, N^{-1/3} + 2.59\, N^{-2/3}\ , \tag{6.8}$$

which gives a value in agreement with that calculated for the infinite-liquid. The rms radii of the N = 3÷112 clusters calculated with the GFMC method are plotted in Fig.6.8. The rms radii increase linearly with $N^{1/3}$ for N > 20, showing that the clusters of N > 20 are like liquid droplets.

The calculated density distributions are shown in Fig.6.9. The density $\rho(N, r=0)$ is shown to decrease as N decreases. This behavior should be

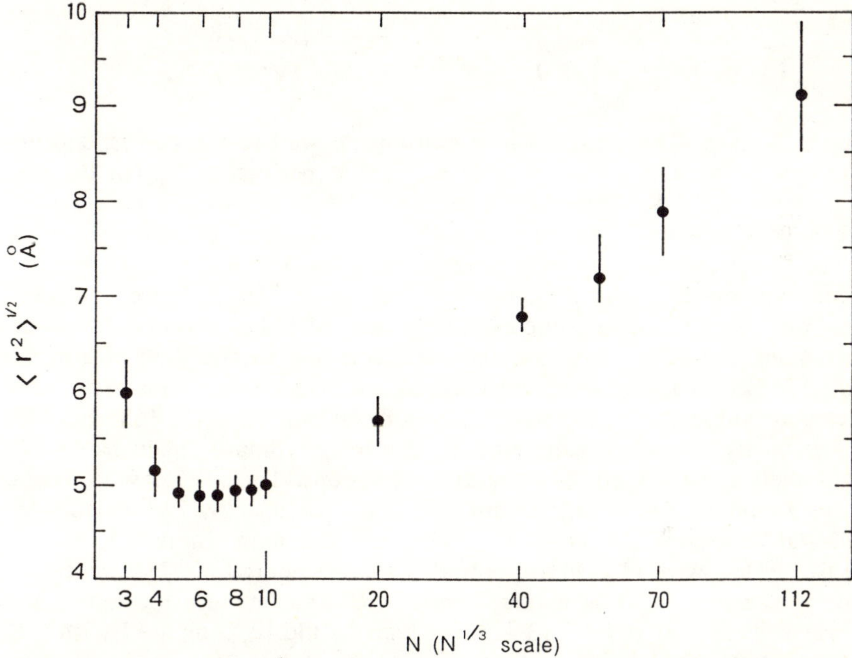

Fig.6.8. The rms radii of ^4He_N clusters calcualted with the GFMC method [6.7]

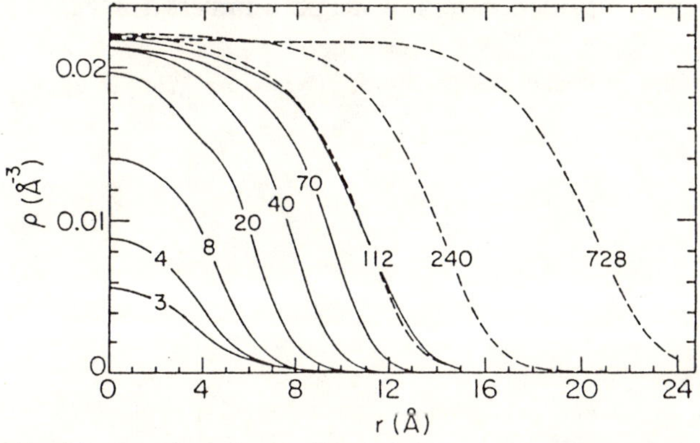

Fig.6.9. Density distributions of ^4He_N calculated with the GFMC (solid lines) and VMC (dashed lines) methods. The curves are labeled with the number of atoms N [6.7]

contrasted with that of a classical liquid droplet whose $\rho(N, r=0)$ should increase as N decreases because of the pressure increase by the surface tension. The calculated density of any cluster does not exceed the liquid density.

The calculated energies of the clusters of $20 \leq N \leq 112$ with the GFMC method and $20 \leq N \leq 728$ with the VMC one can be expressed in the form of an expansion in powers of $N^{-1/3}$

$$E(N)/N = E_v + E_s N^{-1/3} + E_c N^{-2/3} \quad \text{with} \tag{6.9}$$

$$E_v = \begin{cases} -7.02 \\ -6.91 \end{cases}, \quad E_s = \begin{cases} 18.8 \\ 18.9 \end{cases}, \quad E_c = \begin{cases} -11.2 & \text{(GFMC)} \\ -12.0 & \text{(VMC)} \end{cases},$$

in degrees Kelvin. These expressions extrapolate well to the corresponding infinite-liquid energies given in Table 6.2: The experimental E_v for the infinite liquid is -7.12 K. Both the GFMC and VMC values of E_s give a surface tension of the infinite liquid, $\sigma = E_s/[4\pi r_0^2(\infty)]$, of 0.30 $KÅ^{-2}$ which should be compared with the experimental value of 0.27 $KÅ^{-2}$.

Quantum mechanical calculations of clusters of ^3He, a Fermi particle, are much more difficult as compared with those of ^4He_N clusters. To avoid complications caused by intrinsic deformations due to the shell effect, as described in Sect. 3.4, the VMC calculations are made for the clusters of the shell-closing number of the particles, N = 8, 20, 40, 70, 112, 168 and 240 [6.9]. The variational wave functions include the Feynman-Cohen backflow [6.10] as well as two- and three-body correlations. The backflow correlations are found to give a significant effect on the energy. The calculated ground-state energies per atom for ^3He_N are given in Table 6.3, which reveal that ^3He_N with $N \geq 40$ are bound while the energy of $^3He_{20}$ is positive. By expressing E(N)/N in the form of (6.9), the surface tension of liquid ^3He is estimated to be 0.13 $KÅ^{-2}$ which should be compared with the experimental one, 0.11 $KÅ^{-2}$.

In concluding this section, it is emphasized that quantum mechanical calculations do not predict the presence of the magic numbers for ^4He_N.

Table 6.3. The calculated ground-state energies per atom, E(N)/N, for ^3He_N [6.9]

N	E(N)/N [K]
20	+0.21
40	-0.04
70	-0.28
112	-0.46
168	-0.62
240	-0.74
∞	-2.36

This is in harmony with the conclusion mentioned at the end of Sect.6.2.1 that the magic numbers observed for neutral ^4He_N reflect the size distribution realized after the fragmentation in the ionization space.

7. Molecular Clusters

Molecular clusters are aggregates of molecules such as NH_3, H_2O, CO_2, etc. weakly bound by hydrogen bonding or the van der Waals force. When they are charged by photoionization or landing of slow electrons, the electric polarization due to the additional charge plays an important role in binding. Such a situation is quite similar to that of rare-gas clusters mentioned in the previous chapter. Studies of molecular clusters are of much interest from a chemical point of view: Water clusters may be viewed as a microscopic system relevant to solvation. Ammonia clusters and their photochemical products are interesting in intersteller chemistry, as ammonia is an important constituent of the atmospheres of some planets.

7.1 Photoionization of Ammonia Clusters

Ammonia clusters are studied by the molecular-beam-mass spectroscopy with electron-impact ionization and photoionization. The beam is formed by using a conventional supersonic nozzle source. Figure 7.1 shows the observed intensities of the protonated and unprotonated cluster ions of ammonia as a function of the cluster size n with photoionization by Kr resonance line at 10.64 and 10.03 eV [7.1][1]. Being independent of the methods of ionization, it is generally found that most of the prominent peaks in the mass spectra are those of protonated ammonia clusters, $(NH_3)_nH^+$, produced by intracluster ion-molecule reaction.

Small peaks of unprotonated ammonia clusters, $(NH_3)_n{}^+$, are observed in the photoionized mass spectra for any size n up to ~20, as illustrated in Fig.7.1. In the electron-impact mass spectra, these peaks are observable for $n > 10$. The intensities of these peaks are increased when the energies of the impact electrons are lowered.

Detailed examinations lead us to the conclusion that both the protonated and unprotonated clusters have shell structures with an NH_4^+ ion in its center. For examples, $(NH_3)_5H^+$ and $(NH_3)_5^+$ have the structures schematically described as

[1] In this chapter, we use n to denote cluster size instead of N to avoid confusion with N for nitrogen

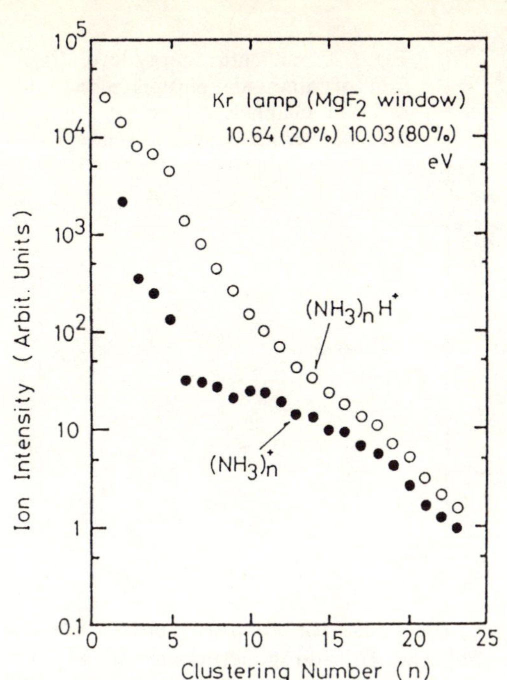

Fig.7.1. The observed intensities of the protonated and unprotonated cluster ions of ammonia as a function of cluster size n with photoionization by the Kr resonance line at 10.64 and 10.03 eV [7.1]

$$\begin{array}{ccc} HNH_3 & & HNH_3 \\ NH_3H - N^+ - HNH_3 & \text{and} & NH_3H - N^+ - HNH_2 \quad . \\ HNH_3 & & HNH_3 \end{array} \tag{7.1}$$

The central molecular ion, NH_4^+, and the shell molecules are bound by the so-called weak hydrogen bond and the polarization due to the central charge. This situation is quite similar to that of rare-gas cluster ions, where the central core ion and the shell atoms are bound by the weak van der Waals force and the polarization due to the core charge, as discussed in Sect. 6.1.2. It is interesting to observe an anomalous intensity drop in the mass spectra at n = 5→6, as revealed in Fig.7.1. This anomaly is certainly related to a saturation of the shell bound with four H atoms of the central molecular ion, NH_4^+, as may be inferred from (7.1).

The relative intensity of $(NH_3)_n^+$ to $(NH_3)_{n-1}H^+$ is determined by competition of the following two reaction paths;

$$(NH_3)_n + h\nu \rightarrow [(NH_3)_n^+]_{vip} + e^-$$

$$\rightarrow \begin{cases} (NH_3)_{n-2}NH_4^+\text{-}NH_2 + e^- & (7.2a) \\ (NH_3)_{n-2}\,NH_4^+ + NH_2 + e^- & (7.2b) \end{cases}$$

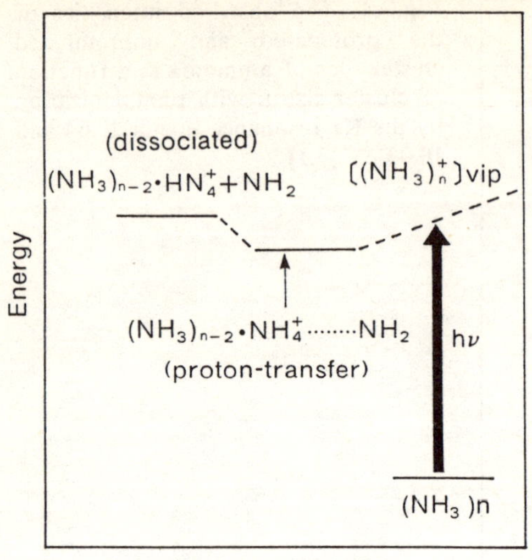

Fig.7.2. Schematic energy level diagram of ammonia clusters along the reaction channels. Subscript vip represents vertically ionized states [7.1]

which are illustrated in Fig.7.2. Here, the subscript vip represents a vertically ionized state. It is assumed that the following sequence of events happens in an ammonia cluster after the near-threshold photoionization: (i) An electron takes off the cluster in 10^{-15} - 10^{-16} s, leaving the "unrelaxed" cluster behind. (ii) Then, proton transfer and reconstruction of the cluster occur in 10^{-13} - 10^{-14} s. (iii) The reconstructed cluster with an NH_4^+ ion at its center is fragmented or boils off NH_2 molecules; it arrives at a thermodynamically stable state in 10^{-5} - 10^{-6} s before reaching the detector of a mass spectrometer.

7.2 Ion-Centered Cage Structure

7.2.1 Magic Numbers for Water Clusters

Experiments similar to those for ammonia cluster beams, as described in the previous section, are also performed for water clusters. Figure 7.3 exhibits a portion of a typical electron-impact mass spectrum of water clusters $(H_2O)_nH^+$ ($17 \leq n \leq 32$) with the ionization electron energy of 40 eV [7.2]. A distinct intensity drop between n = 21 and 22 can be seen, so that n = 21 may be called a magic number. Experimental evidence for the existence of the magic number stability in water clusters was first reported in 1973 [7.3]. Then, it has been proposed that a clathrate-like pentagonal dodecahedron with an H_3O^+ ion at the center of the cage, as depicted in Fig.7.4, may be the stable structure of $(H_2O)_{21}H^+$ [7.4, 2]. The structure contains a kernel structure, $(H_2O)_3H_3O^+$, which is the smallest structure with the shell molecules hydrogen-bonded with the core ion H_3O^+. This shell structure corresponds to that of $(NH_3)_4NH_4^+$ in (7.1).

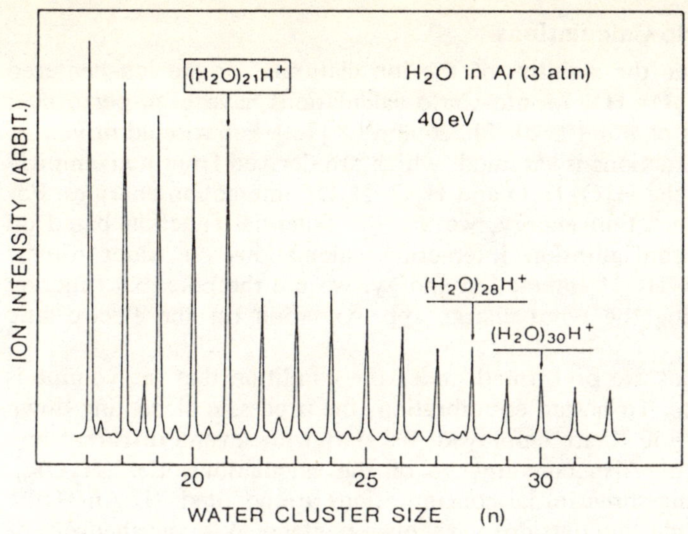

Fig.7.3. Electron-impact (40eV) mass spectrum of water clusters $(H_2O)_nH^+$ in the size range of n = 17-32. Magic-number ions are indicated by arrows [7.2]

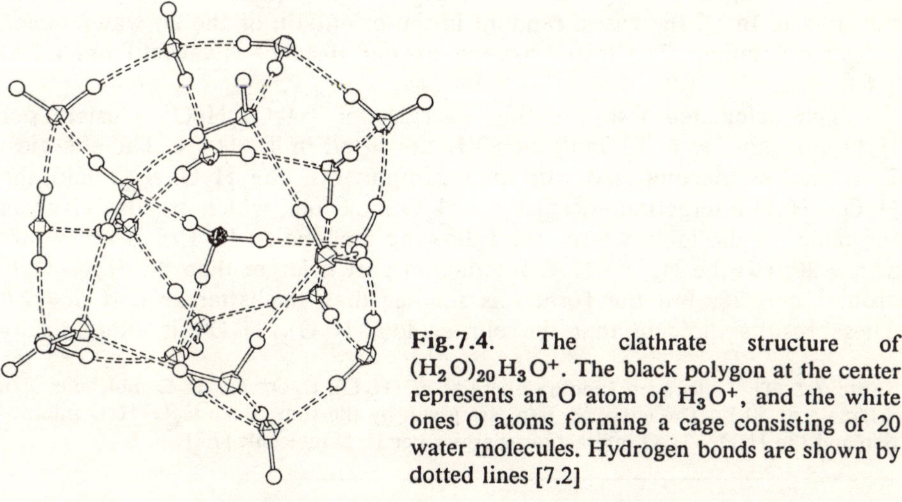

Fig.7.4. The clathrate structure of $(H_2O)_{20}H_3O^+$. The black polygon at the center represents an O atom of H_3O^+, and the white ones O atoms forming a cage consisting of 20 water molecules. Hydrogen bonds are shown by dotted lines [7.2]

Figure 7.3 also shows less distinct irregularities at n = 28 and 30. These anomalies are not always clearly observed, depending upon the beam expansion conditions. However, as in the case of n = 21, intensity enhancements are observed at n = 28 and 30 in the free jet expansion of ionized water vapor as well as in the ionization of neutral water clusters.

The intensity drop between $(H_2O)_{21}H^+$ and $(H_2O)_{22}H^+$ is also observed in the photoionization mass spectrum by using a resonance line emission of an Ar resonance lamp at 11.82 eV.

7.2.2 Monte-Carlo Calculations

In order to examine the stability of the ion clathrate, or the ion-centered cage model for $(H_2O)_nH^+$, Monte-Carlo calculations have been performed for the cluster ions of n = 19, 20, 21, 26 and 28 [7.2]. Pairwise additivity of intermolecular interactions is assumed, which are derived from non-empirical calculations of the H_2O-H_2O and H_2O-H_3O^+ interaction energies. For the H_2O-H_2O interaction energy, we use the potential function based on moderately large configuration interaction calculations on water dimers [7.5]. For the H_2O-H_3O^+ interaction energy, we use the potential function derived by assuming the point charge approximation for the electrostatic part [7.6].

The calculations are preformed under the condition that the volume is allowed to fluctuate. To hasten equilibration, the processes of cooling down as 200→150→100→50 K are employed. We start with several different initial conditions to confirm convergence of the calculations. For $(H_2O)_{20}\cdot H_3O^+$, the following three initial configurations are adopted: (i) An H_3O^+ ion is placed inside or outside a regular pentagonal dodecahedron of $(H_2O)_{20}$; (ii) an H_3O^+ ion is placed at the center of a body-centered cubic structure; (iii) the oxygen atoms of 20 water molecules are distributed randomly on a spherical surface and an H_3O^+ ion is placed inside or outside the sphere. In all the cases, random initial orientatin of the 20 water molecules are assumed. The initial oxygen-oxygen distance is varied from 1.7 to 3.4 Å.

The calculated total binding energies of $(H_2O)_nH_3O^+$ clusters per H_2O molecule, E/n (kJ/mol), at 50 K are listed in Table 7.1. The energies, E/n, can be decomposed into two components, the H_2O-H_2O and the H_3O^+-H_2O interaction energies per H_2O molecule, which are also given in the table. In the table we see the following tendencies: (i) E/n is maximum at n = 20; (ii) the H_3O^+-H_2O interaction part is larger than the H_2O-H_2O around n = 20, but the former is smaller than the latter around n = 27. These results indicate that the cluster ion $(H_2O)_{20}H_3O^+$ is energetically

Table 7.1. The calculated binding energies of $(H_2O)_nH_3O^+$ per H_2O molecule, E/n [kJ/mol], at 50 K. The energies, E/n, are given by the sum of the H_2O-H_2O interaction and the H_3O^+-H_2O interaction energies per H_2O molecule [7.2]

n	E/n[a]	H_2O-H_2O int.	H_3O^+-H_2O int.
19	48.23	21.64	26.59
20	49.66	24.04	25.62
21	47.92	21.80	26.12
26	47.99	28.28	19.71
27	47.56	26.55	21.01
28	46.70	28.74	17.96

[a] The values given in [7.2] are E/(n+1)

stable and the H_3O^+-H_2O interaction energy is comparable to the H_2O-H_2O.

7.3 Negatively Charged Water Clusters

7.3.1 Solvated Electrons

It is well known that electrons can be trapped in localized states in liquid or solid water although a water molecule does not have a stable negatively charged state. Numerous studies have been performed on the behavior of low-energy electrons in water vapor and liquid water. It has turned out that the ground state of an excess electron in liquid water is about 1-2 eV below the vacuum level. The state corresponds to a solvated electron self-trapped: water dipoles in the first coordination shell are oriented toward the excess electron at the center.

A sufficiently large water cluster also solvates an electron. It has been suggested that more than eight water molecules are required to get a cluster anion with the ground state stable against autodetachment [7.7]. The adiabatic electron affinity of $(H_2O)_n$ is close to zero until it starts to converge toward the bulk value.

When electrons of 6-14 eV energies are incident to the beam of neutral water clusters, the main products are found to be $(H_2O)_nOH^-$. No negatively charged clusters, $(H_2O)_n^-$, are observed [7.8]. According to a theoretical calculation [7.9] for $(H_2O)_nOH^-$, an electron is localized at OH having a large electron affinity (1.8eV). Surrounding OH^-, water molecules form a shell. This situation is similar to that found for $(H_2O)_nH_3O^+$ already discussed in Sect.7.2, but quite different from that for a solvated electron.

7.3.2 Trapping of Electrons

Negatively charged water clusters $(H_2O)_n^-$ ($n \geq 2$, $n \neq 4$) are observed by injecting low-energy electrons into a supersonic expansion of water seeded in rare gas [7.10]. Figure 7.5 exhibits a mass spectrum of negative ions obtained from expansion of 2% H_2O in Ar. Peaks corresponding to $(H_2O)_n^-$ with n = 2, 6, 7 and $\geq$10 are clearly observed. Only the peaks for $n \geq 10$ are observed in pure H_2O expansions or when the H_2O partial pressure in rare gases is larger than about 10%.

The primary effect of seeding in rare gases is to cool the clusters, which enhances the formation of small $(H_2O)_n^-$. The presence of weakly bound $Ar(H_2O)_n^-$ seen in Fig.7.5 confirms that the clusters are indeed cold. At higher H_2O partial pressures, the effective temperature of the cluster is increased, which prevents the formation of small $(H_2O)_n^-$ and $Ar(H_2O)_n^-$ clusters. It is interesting to observe $(H_2O)_2^-$ whose adiabatic electron affinity is estimated to be 17 meV from electric-field-detachment experiments [7.12].

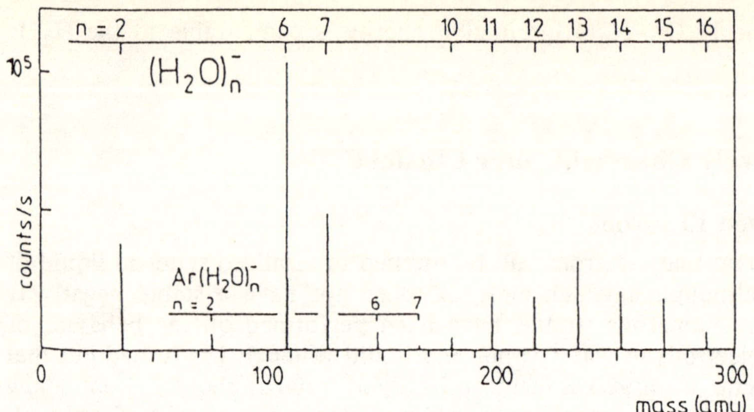

Fig.7.5. Mass spectrum obtained by injecting slow electrons into a supersonic expansion of water seeded in argon [7.10]

An entirely different method is used to obtain long-lived water cluster anions. The method employs attachment of very-low-energy electrons to preformed cold water clusters in a well collimated cluster beam [7.11]. No collisions are required to stabilize the charged clusters in contrast to the case of using seeding in rare gases [7.10]. Most experiments are run with deuterated water in order to facilitate identification of the anion composition. The supersonic expansion jet is skimmed, collimated, and intersected by a pulsed electron beam. The electrons are emitted from a directly heated filament. The width of the energy distribution is about 1 eV.

For electron energies around 7 eV, cluster anions with the composition $(D_2O)_{n-1}OD^-$ ($n \geq 2$) are observed, as shown in the top of Fig.7.6 [7.11]. The spectrum is the same as observed in a similar experimental setup in [7.8]. Below 1 eV, cluster anions $(D_2O)_n^-$ are observed, as indicated in the bottom of Fig.7.6. Beyond a threshold size of n = 12, the intensity increases sharply and forms a maximum around n = 20. Dependence of the $(D_2O)_n^-$ intensity upon the electron energy near zero eV is measured by comparing with the measured yield curve of SF_6^- and SF_5^-, as shown in the top of Fig.7.7, both being produced by electron attachment to the SF_6 molecule. The position of the $(D_2O)_n^-$ is almost the same as that of SF_6^- within the experimental accuracy. Thus, pure water cluster anions are formed resonantly by electron attachment for energies at or very close to 0 eV with an uncertainty of 0.2 eV. The intrinsic width of this resonance may be very small: It is certainly no wider than 0.5 eV. Electron attached cluster anions may be considered to be in the metastable state.

They can be relaxed to the ground state by evaporating water molecules as

$$(D_2O)_n^- \rightarrow (D_2O)_{n-1}^- + D_2O \, , \tag{7.3}$$

and/or by emitting electrons as

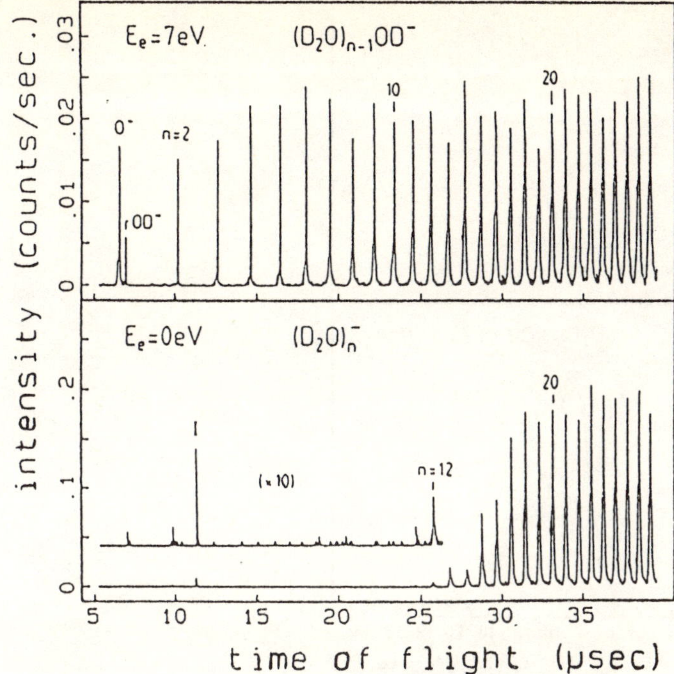

Fig.7.6. Mass spectra of deuterated water cluster anions for 7 eV and nearly 0 eV incident electron energy. "I" denotes an impurity peak [7.11]

$$(D_2O)_n^- \rightarrow (D_2O)_n + e^- \; . \tag{7.4}$$

It is experimentally verified [7.13] that the process of (7.3) occurs with a probability of ~20% for cluster sizes $14 \leq n \leq 20$, and that the electron detachment process of (7.4) does at least with a probability of 25% for the smallest cluster anions ($12 \leq n \leq 14$); it decreases rapidly with increasing cluster size.

7.3.3 Theoretical Treatments

Electron distributions and energies of water cluster anions are theoretically investigated by using Quantum Path-Integral Molecular-Dynamics simulations (QPIMD) [7.14], which is based on the path-integral formulation of quantum statistical mechanics and isomorphism between the quantum problem and a classical one [7.15, 16]. In modeling the system, we use the RWK2-M model [7.17] for the intramolecular and intermolecular interactions. For the electron-water interaction we use a pseudo-potential including the Coulomb, polarization, exclusion and exchange interactions, derived from the density-functional theory.

The energetics of the system can be expressed in terms of the vertical electron affinity (the energy required to detach an electron from the ground state of $(H_2O)_n^-$ without changing the atomic framework) and the adiabatic

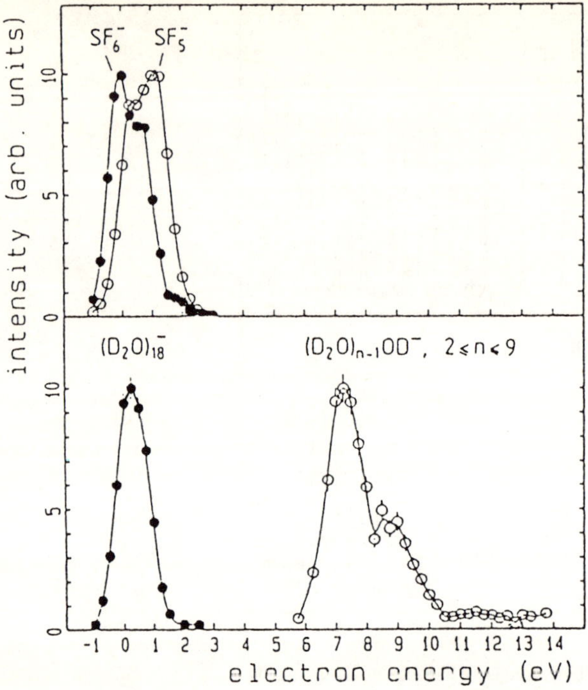

Fig.7.7. Bottom: Energy dependences of the $(D_2O)_{18}^-$ and $(D_2O)_{n-1}$ OD^- intensities upon the energy of an attaching electron. The intensity of $(D_2O)_{n-1}OD^-$ is summed up over several cluster sizes. For a given cluster size, it is only ~8% of that for $(D_2O)_{18}^-$. Top: Yield curves of SF_6^- and SF_5^- produced from SF_6, recorded for the purpose of electron energy calibration [7.13]

electron affinity (the ground state energy of $(H_2O)_n$ minus that of $(H_2O)_n^-$). The $(H_2O)_n^-$ is stable when the adiabatic electron affinity is positive.

The calculated vertical electron affinities (full symbols) and the adiabatic electron affinities (empty symbols) [7.15] are plotted in Fig.7.8 for the two kinds of states, the interior (circles) and the surface (squares) states, where an excess electron is localized in the interior and at the outside (or the surface) of the cluster, respectively [7.16, 18]. In Fig.7.8 we observe that, although the vertical electron affinities are positive and of the order of one eV, they are almost cancelled by the energy gain obtained by reconstructing the atomic framework resulting in the small values of the adiabatic electron affinities of the order of 0.1 eV. The adiabatic electron affinity of the surface state is positive and larger than that of the interior at each n of $8 < n < 64$, but the adiabatic electron affinity of the interior state exceeds that of the surface state for $n \geq 64$. This result indicates that the surface states are stable for small $(H_2O)_n^-$ but the interior states are more stable for larger ones with $n \geq 64$.

The calculated distributions of an excess electron in the surface states of $(H_2O)_2^-$ and $(H_2O)_{64}^-$ and in the interior states of $(H_2O)_{64}^-$ and $(H_2O)_{128}^-$

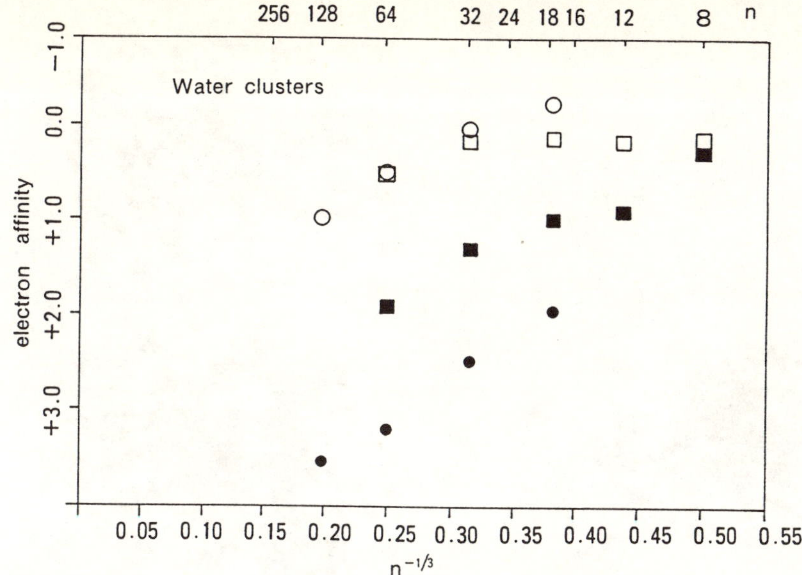

Fig.7.8. The calculated vertical electron affinites (full symbols) and the adiabatic electron affinities (empty symbols) for the interior (circles) and the surface (squares) states of $(H_2O)_n^-$. The effective temperatures are 300 K for $n \geq 32$ and 79 K for $n < 32$ [7.15]

are shown in Fig.7.9 [7.18]. The results of the energetical investigation together with those of the electron distribution are in agreeement with the picture of solvated electrons and also with the finding of a stable $(H_2O)_2^-$ [7.12]. According to the results of the QPIMD studies the stable $(H_2O)_2^-$ has to be in the surface state, which should be verified experimentally.

7.4 Electron Attachment to van der Waals Clusters

7.4.1 Electron Transfer from High-Rydberg Rare-Gas Atoms

A weakly bound outermost electron of a high-Rydberg atom can be collisionally transfered to a molecule having a positive electron affinity with the cross section of the order of 10^{-12} cm^2 [7.19]. Similarly, an efficient production of negative cluster ions is achieved by use of the collisional electron transfer from a high-Rydberg rare-gas atom A^{**} (A = Ar,Kr) to neutral clusters of isoelectronic linear molecules, CO_2, OCS and CS_2 [7.20]. The clusters selected have such properties that each component molecule in the linear configuration posseses a negative vertical electron affinity but the energy gain of 2÷3 eV is expected when the negatively charged molecule is bent.

The cluster beam is formed by free-jet expansion of the sample gas seeded with either hydrogen or helium. High-Rydberg rare-gas atoms A^{**}

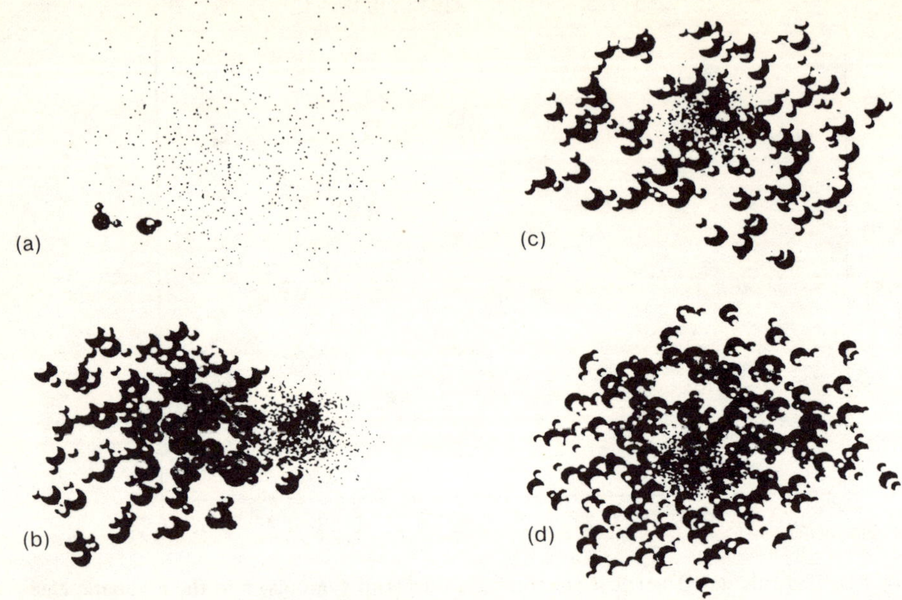

Fig.7.9. The calculated distribution of an excess electron in **(a)** the surface state of $(H_2O)_2^-$; **(b)** the surface state of $(H_2O)_{64}^-$; **(c)** the interior state of $(H_2O)_{64}^-$; **(d)** the interior state of $(H_2O)_{128}^-$ [7.18]

with the principal quantum numbers of 20÷30 are formed by electron impact on the rare-gas atoms, and then allowed to interact with the cluster beam. Figure 7.10 displays the observed mass spectra of negatively charged cluster ions $(M)_n^-$ (M = CO_2, OCS and CS_2) produced by the collision of Kr^{**}. The spectra show the following common features: a) the spectra obtained by using different rare gases are almost identical; b) the intensity rises at a threshold size, n_L, increases with increasing n, passes through a broad maximum at about n_L+10, and falls off gradually (n_L = 7,2,1 for CO_2, OCS, CS_2, respectively); c) several outstanding peaks are discernible at n_M which may be called the magic numbers; d) a depletion region ($11 \leq n \leq 13$) is observed for $(CO_2)_n^-$.

7.4.2 The Strongly-Coupled Electron-Phonon Model

In order to describe the electron attachment process, as mentioned in the previous subsection, simple model Hamiltonian is used as follows [7.21]

$$\mathcal{H} = \sum_{\mathbf{k}} \epsilon(\mathbf{k}) a_{\mathbf{k}}^{+} a_{\mathbf{k}} + \sum_{i} \left\{ \epsilon_i + \sum_{\lambda} \eta_{\lambda}^{i} (b_{\lambda}^{+} + b_{\lambda}) \right\} a_i^{+} a_i + \sum_{i,\mathbf{k}} \left[\Theta_i(\mathbf{k}) a_i^{+} a_{\mathbf{k}} + \text{h.c.} \right] + \sum_{\lambda} \hbar\omega_{\lambda} b_{\lambda}^{+} b_{\lambda} , \quad (7.5)$$

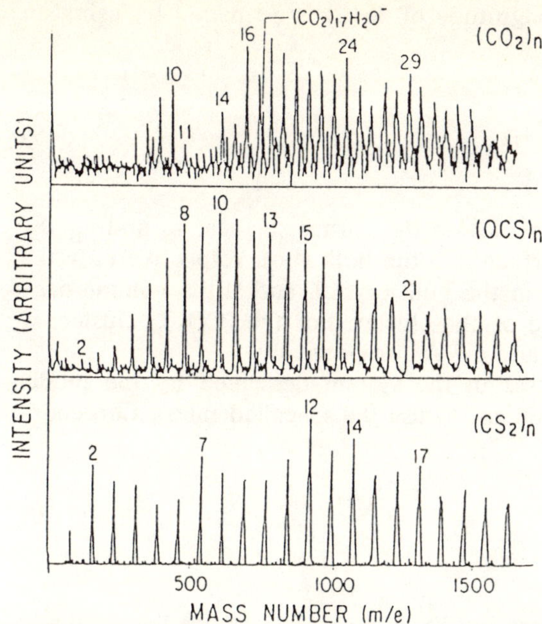

Fig.7.10. Mass spectra of negative cluster ions produced by collision of a high-Rydberg Kr atom with neutral clusters $(M)_n$ ($M = CO_2$, OCS and CS_2) [7.20]

where $a_{\mathbf{k}}$ and a_i are the electron annihilation operators for the free state **k** and the affinity state i of the cluster, respectively. Energy ϵ_i is the affinity level position of the neutral cluster with the equilibrium geometry. Operators b_λ and b_λ^+ are the boson annihilation and creation operators corresponding to the λth normal mode of vibrations. The second term of (7.5) takes into account the dependence of the affinity levels $E_i(\{\mathbf{R}_j\})$ on the cluster deformation $\mathbf{R}_j - \mathbf{R}_j^o$ up to its first order, so that

$$\epsilon_i = E_i(\{\mathbf{R}_j^o\}) \, , \tag{7.6}$$

$$\eta_\lambda^i = \sum_{j\alpha} C_{\lambda j\alpha}^* \sqrt{\frac{\hbar}{2\omega_\lambda}} \, \nabla_{j\alpha} E(\{\mathbf{R}_j^o\}) \quad (\alpha = x, y, z) \, . \tag{7.7}$$

Here, $\{\mathbf{R}_j^o\}$ is the equilibrium geometry and $C_{\lambda j\alpha}$ the coefficients of the linear combination giving the normal coordinate ξ_λ as

$$\xi_\lambda = \sum_{j\alpha} C_{\lambda j\alpha}(\mathbf{R}_{j\alpha} - \mathbf{R}_{j\alpha}^o) \, . \tag{7.8}$$

In (7.5), the non-vanishing coupling constant $\Theta_i(\mathbf{k})$ between the free state **k** and the affinity state i arises mainly from the presence of surface molecules in the cluster. The magnitude of $\Theta_i(\mathbf{k})$ is estimated by using an approximate relation,

$$|\Theta(\mathbf{k})|^2 \sim \frac{n_s}{n}\left(\frac{z_B - z_S}{2z_B}\right)^2 w^2\Omega\,, \tag{7.9}$$

where n_s is the number of molecules on the cluster surface, z_S and z_B the coordination numbers of the surface and the bulk sites, respectively, w the bandwidth of the affinity level in the bulk crystal, and Ω the volume occupied by a component molecule of the cluster. For the $(CO_2)_n$ cluster, w and Ω are assumed to be 2.1 eV and 6.7 Å^3, respectively.

To discuss physical processes in the system described by the model Hamiltonian of (7.5), it is convenient to use the so-called interaction coordinate given by

$$Q^i \equiv -\sum_{\lambda} \eta_\lambda^i (b_\lambda^+ + b_\lambda)\,, \tag{7.10}$$

in place of the normal coordinates in (7.8). Figure 7.11 shows the adiabatic potential $V^{\mathbf{k}}$ of the system consisting of a neutral cluster and a non-interacting free electron in the **k** state and the adiabatic potential V^i of the negatively cahrged cluster as a function of the interaction coordinate Q^i. In the systems of interest, Q^i is approxiamtely considered to be the coordinate of the bending mode of a component molecule of the cluster. The figure indicates that the bending is absent in the ground state of the neutral cluster but it stabilizes the negatively charged cluster as shown by V^i.

The process of the electron attachment develops as follows: i) The attachment of an electron excites the system to the vertical affinity state ϵ_i, whose electron orbital spreads over the whole cluster; ii) the excited system

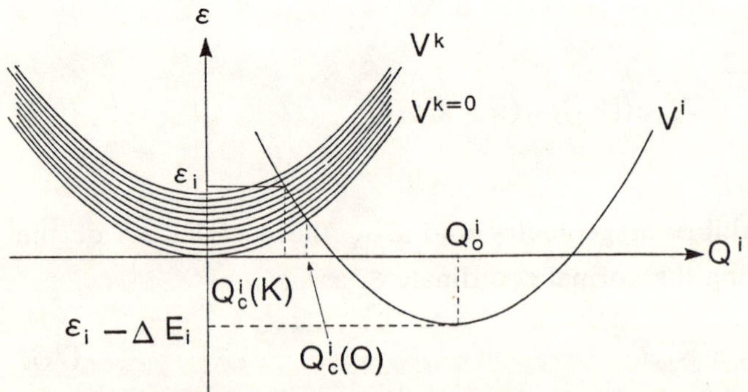

Fig.7.11. Adiabatic poteridal $V^{\mathbf{k}}$ of the system-consisting of a neutral cluster and a non-interacting free electron in the **k** state and adiabatic potential V^i of the negatively charged cluster as a function of the interaction coordinate Q^i. ϵ_i is the value of ϵ where $V^{\mathbf{k}}$ crosses V^i [7.21]

may decompose into a free electron in the **k** state and the neutral cluster with the bending displacement $Q_c^i(\mathbf{k})$ until Q^i reaches $Q_c^i(0)$ or may arrive at the stable equilibrium state of the negatively charged cluster with the bending Q_0^i and an additional electron localized around the bent component molecule. The stable reconstructed cluster anion is realized if the vertically excited system survives without reemitting an electron for the time longer than τ_c^i: τ_c^i is the time needed for the system to move from $Q_c^i(\mathbf{k})$ to $Q_c^i(0)$ along the curve V^i.

By using the diagrammatic expansion method [7.22], the cross section of the electron attachment is obtained as

$$\sigma \sim \frac{2\pi}{\hbar v} \sum_i |\Theta_i(\mathbf{k})|^2 \frac{1}{2\sqrt{\pi}\Gamma_i} \exp\left(\frac{-(\epsilon(\mathbf{k})-\epsilon_i)^2}{4\Gamma_i^2}\right) \exp\left(-\frac{2\pi}{\hbar}\langle|\Theta_i(\mathbf{k})|^2\rangle \rho_{\mathbf{k}} \tau_c^i\right) \qquad (7.11)$$

where $\rho_{\mathbf{k}}$ and v are the density of states and the velocity of the **k** free electron, respectively, and Γ_i the width of the vertical affinity level due to vibrations. The last exponential factor of (7.11), where

$$\tau^i \equiv \left[-\frac{2\pi}{\hbar}\langle|\Theta_i(\mathbf{k})|^2\rangle \rho_{\mathbf{k}}\right]^{-1} \qquad (7.12)$$

represents the life time of the vertical affinity state i, gives the survival probability of the ith affinity state during time τ_c^i, and the preexponential factor the transition probability of the system being excited to the ith affinity state by a free electron in the **k** state.

To estimate the value of τ^i, it is most important to know the distribution of the vertical affinity levels ϵ_i. It is calculated by using a non-empirical so-called DV-Xα-transition state method [7.23]. The result for $(CO_2)_n$ is shown in Fig.7.12, where we observe a systematic trend of lowering of the level positions with the increase of cluster size. The level positions relative to the vacuum level E_{vac} are slightly dependent upon the atomic basis set used in the calculation, although the relative positions of clusters of different size are insensitive to the basis set. In the figure, the vacuum level is located 4.0 eV below the affinity level of the monomer in accord with the result of the Hartree-Fock-CI calculation [7.24].

By using the approximate relation in (7.9) for $|\Theta_i(\mathbf{k})|^2$, and appropriate values for Γ_i and τ_c^i, the calculated cross section σ for the electron attachment on $(CO_2)_n$ is presented in Fig.7.13 as a function of the electron incident energy ϵ. The figure shows that the cross section is negligibly small for small clusters of $n = 2$ and 4, but takes significant values for $n \geq 7$. This is in fair agreement with the experimental result depicted in Fig.7.10. If the affinity levels are located at higher energies, the survival probability given by the last exponential factor of (7.11) becomes vanishingly small. This is the main reason why small clusters of $n \leq 4$ cannot capture an electron.

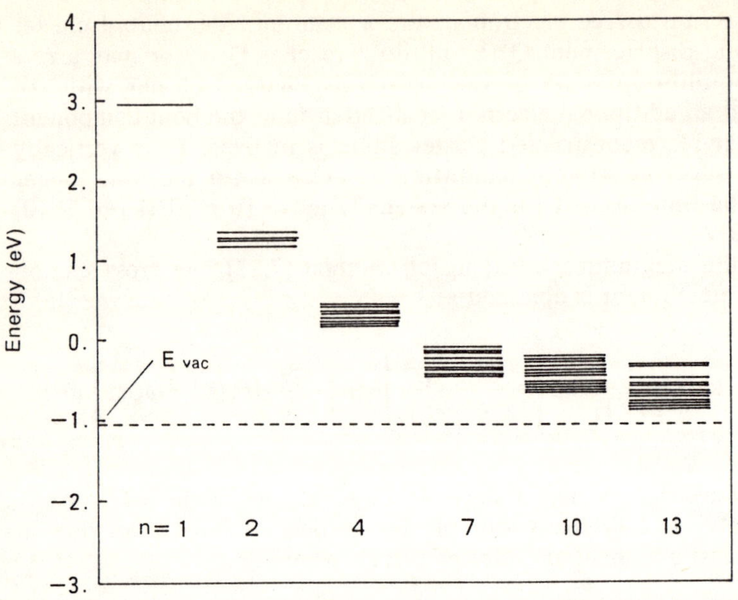

Fig.7.12. The calculated distribution of the vertical affinity levels of $(CO_2)_n$ [7.21]

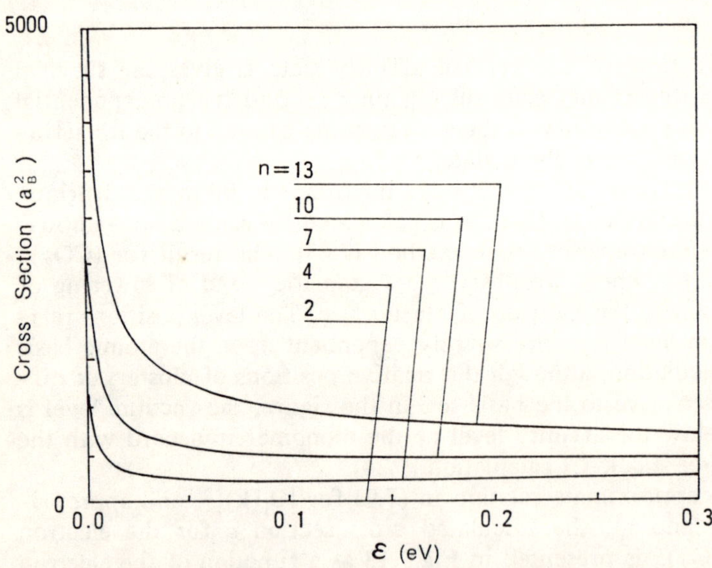

Fig.7.13. The estimated cross section of the attachment of an electron with kinetic energy ϵ for $(CO_2)_n$ [7.21]

Theoretical understanding of heating and evaporation processes of the clusters after the electron attachment is very important in explaining the observed fine structure of mass spectra, as seen in Fig.7.10. This problem is left to be studied in the future.

8. Miscellaneous Topics

Some recent topics in the field of microcluster physics and chemistry, not touched upon in the preceding chapters, are arbitrarily picked out and briefly discussed in this last chapter.

8.1 Synthetic Chemistry in a Cluster Beam

Studies of chemical reactions in the gas phase are very attractive, as one can avoid complications arising from solvents or matrix effects. An apparatus is developed to allow us to perform the above-mentioned studies for naked metal clusters of a specific size and charge, as shown in Fig.8.1 [8.1]. A sputtering arrangement is chosen as the cluster source, which generates positively and negatively charged clusters of various size and charge as well as neutrals from nearly all kinds of materials. The emitted cluster ions are energy-selected, mass-separated, and then introduced into an ion drift tube where they are slowed down to almost thermal velocities giving residence times of up to 10 ms in a radio-frequency confinement. The confined cluster ions are exposed to physical perturbations or allowed to react with other species. The product ions are then analyzed with another mass spectrometer at the exit of the ion drift tube.

When a nickel target is used, cluster ions Ni_N^+ of size $N = 1 \div 20$ are produced. Nickel cluster ions of any fixed size are selected by mass separation. Introduction of the cluster ions Ni_N^+ into the ion drift tube with a low

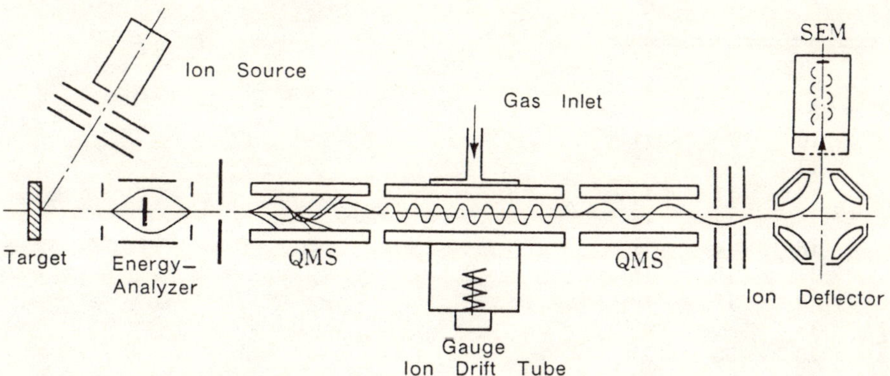

Fig.8.1. Experimental setup for studying physical and chemical properties of clusters of a fixed size in a gas phase [8.1]

pressure of carbon monoxide and subsequent mass spectrometric analysis of the products reveals the formation of three serieses of $Ni_N(CO)_k^+$, $Ni_NC(CO)_l^+$, and $Ni_{N-1}(CO)_m^+$ cluster ions. Saturation of the cluster with carbon monoxide ligands occurs when the pressure of CO is gradually increased up to the point at which the product spectrum does not change. The saturated spectra observed for N = 4, 6, and 10 are shown in Fig.8.2. The maximum numbers of CO ligands for the three serieses of nickel clusters of N = 2÷13 are listed in Table 8.1.

It is interesting to compare the results in Table 8.1 with the work of *Lauher* [8.2] which discusses the most favorable molecular geometry for any given transition metal cluster predicting its bonding capabilities. For example, a tetrahedral metal cluster is predicted to be stabilized when the total number of valence electrons of the cluster is 60. This total is made up of metal valence electrons and ligand valence electrons: the number of ligand valence electrons is two from each carbon monoxide. Hence, $Ni_4(CO)_{10}^+$ ion will have a stable tetrahedral arrangement of nickel atoms. A similar argument is applied to the larger ligated nickel clusters listed in Table 8.1 [8.1].

8.2 Latent-Image Generation

It is well known that the photographic process in light-sensitive silver-halide microcrystals develops in the following way. Absorption of a photon creates a conduction electron and a positive hole. Under favorable conditions the conduction electron may combine with a mobile interstitial silver ion giving an isolated silver atom. Repetition of such a process forms a

Table 8.1. Maximum numbers of CO ligands as a function of cluster size for three serieses of reaction channels; $Ni_N^+ + CO \rightarrow Ni_N(CO)_k^+, \rightarrow Ni_NC(CO)_\ell^+, \rightarrow Ni_{N-1}(CO)_m^+$ [8.1]

N	k	ℓ	m
2	9	5	8
3	8	7	9
4	10	7	11
5	12	9	13
6	13	11	11
7	15	13	14
8	16	14	17
9	17	12	18
10	18	16	19
11	19	-	20
12	20	20	21
13	22	20	22

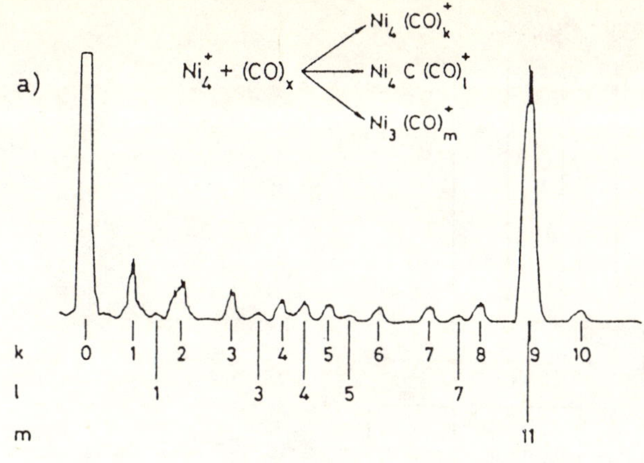

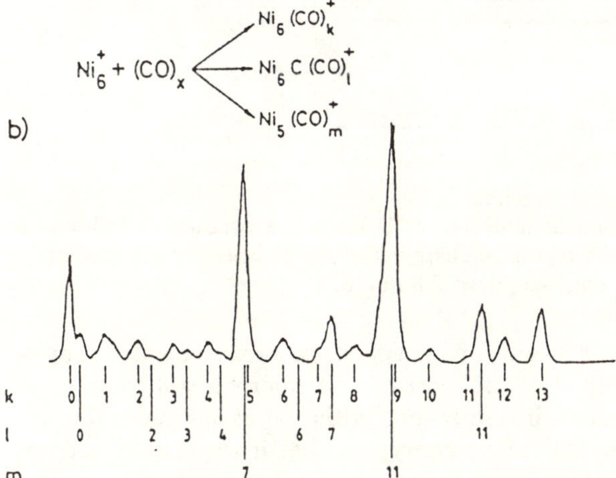

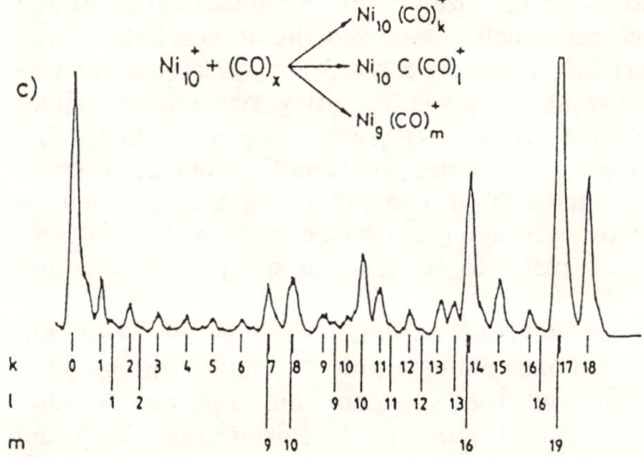

Fig.8.2. Products of the reaction of carbon monoxide with (a) Ni_4^+, (b) Ni_6^+, and (c) Ni_{10}^+ at a CO pressure of approximately $3 \cdot 10^{-3}$ mbar [8.1]

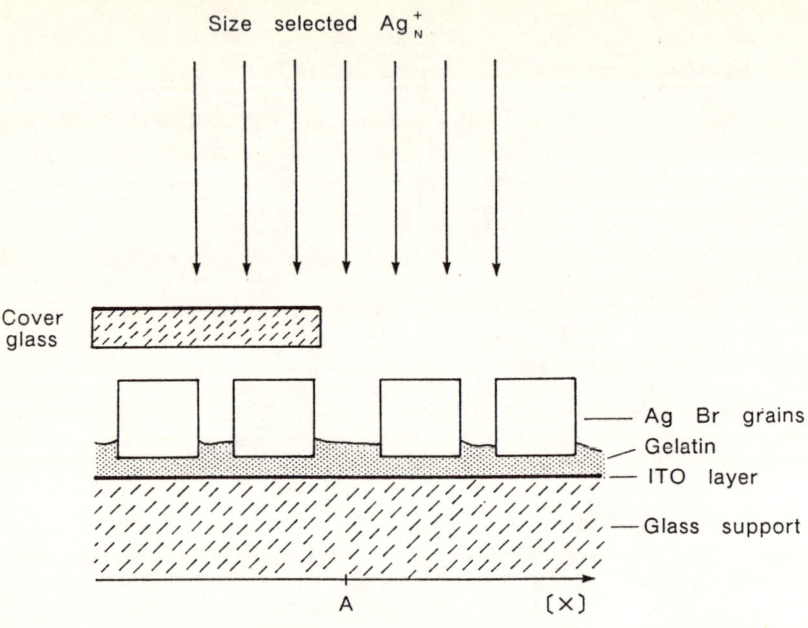

Fig.8.3. Schematic specimen arrangement. The ITO layer is a conductive indium-tin oxide layer to avoid build-up of a repulsive charge. The monodisperse AgBr microcrystals are of cubic shape with an edge length of 0.8 μm [8.3]

cluster of silver atoms called a latent-image speck. Its presence subsequently accelerates or catalyzes complete reduction of the microcrystal to metallic silver. In the absence of a latent-image speck, initiation of the reduction requires a considerably higher activation energy so that it starts after a comparatively long induction time.

It has been one of the most important problems to answer the question how large is the critical latent-image size. Within a limited range of the redox potential of the developer usually employed, numerous indirect investigations show that the critical size is around a few silver atoms. An unambiguous direct proof, however, is given by using size-selected silver clusters to create mono-disperse latent-image silver aggregates [8.3]. The experiment for the direct proof is performed, replacing the blocks after the first Quadrupole Mass Spectrometer (QMS) in Fig.8.1 by the specimen, as depicted in Fig.8.3. Half of the impinging cluster beam is shaded off with an electrically conductive glass plate, to account for spurious background radiation.

After exposure to the size-selected cluster beam, the specimens are developed for 30 s in a conventional photographic developer without subsequent fixation. The developed and dried specimens are examined by using an optical microscope. The observed fraction of developed grains is plotted versus the counting coordinate X in Fig.8.4. The specimen area shielded by the cover glass is at the left-hand side of the point marked A. In the figure

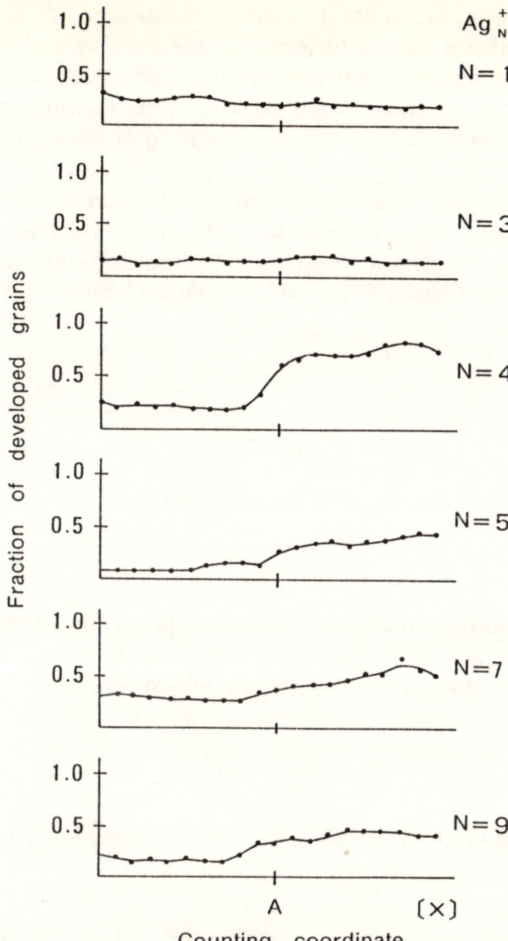

Fig.8.4. Fraction of developed grains versus counting coordinate X, which, together with point A, is shown in Fig.8.3 [8.3]

it can clearly be seen that no development is obtained for the exposure to the beams of Ag_1^+ and Ag_3^+. The exposure to the Ag_4^+, however, causes a remarkable increase of developability to nearly 80%. Similarly, some increase of the developability is also observed for larger clusters, Ag_5^+, Ag_7^+ and Ag_9^+. These results lead us to the conclusion that development requires a critical size aggregate of four silver atoms within a limited range of the redox potential of the developer usually employed.

8.3 Electron Correlation and Chemical Bonds

The purpose of this section is to point out the importance of effects of electron correlations in determining the geometry of microclusters. For small molecules the so-called configuration-interaction method is employed to take into account electron correlations to great accuracy. However, it requires too much computation time for microclusters of large size. A technique successfully applied to large systems takes into account self-consistent potential for a gas of free electrons with the same density. This technique called the Local-Density-Functional (LDF) method was described in Sect.3.2.1.

The LDF method fails to deal with the case where the electron repulsion is large, as comapred with the kinetic energy. In such a case, we have to simplify the description of the electronic properties by making atomic averages, i.e. by using a Linear Combination of Atomic Orbitals $|a\rangle$ (LCAO) [8.4],

$$\psi = \sum_{a} \alpha_a |a\rangle \ . \tag{8.1}$$

To describe the kinetic energy we use interatomic transfer integrals

$$t = \langle a | v_a | b \rangle \ , \quad (t < 0) \tag{8.2}$$

where v_a is the potential energy acting on an electron in the $|a\rangle$ orbital, b is usually limited to the nearest neighbors of a.

In what follows, we adopt a simplified model, in which the atomic orbitals are orthogonal,

$$S = \langle a | b \rangle = 0 \ , \tag{8.3}$$

and only a non-vanishing integral of the electron-electron interaction v_{12} is the intraatomic repulsion U,

$$U = \langle a(1)a(2) | v_{12} | a(1)a(2) \rangle \ . \tag{8.4}$$

This model is called the *Hubbard model* [8.5]. In the Hartree-Fock approximation, the independent electrons are considered to be distributed randomly over all atomic spin orbitals so as to respect locally the Pauli exclusion principle for electrons with the same spin. The effect of retaining the positive U is to reduce the charge fluctuations (even for the electrons of the opposite spin!) produced on each atom by such a random distribution.

If U is much smaller as compared with t, an expansion of the total energy in powers of $|U/t|$ is possible. The first term linear in U is the Coulomb repulsive term computed in the Hartree Fock limit for uncorrelated delocalized electrons. The second term in U^2 gives the first deviation from the Hartree Fock random distribution of electrons, called the effect of electron correlations. Following the arguments by *Friedel* [8.4], we shall point out the importance of electron correlations in the chemical bonds of clusters in this section.

8.3.1 Dimers Versus Close-Packed Aggregates

Let us first calculate the energy of a H_2 dimer, restricting ourselves to a subspace spanned by the 1s atomic orbitals. The energy E_2 of the ground singlet state in the expansion in powers of $|U/t|$ is given by $(t<0)$

$$\begin{aligned} E_2/2 &= E_0 + R/2 + (U - \sqrt{U^2 + 16t^2}) \\ &\simeq E_0 + R/2 + t + U/4 + U^2/(32t) + O(U^3/t^2) , \end{aligned} \tag{8.5}$$

where E_0 is the energy of a neutral H atom, and R the repulsion between two H atoms. The fourth term in the last expression of (8.5), linear in U, is obtained in the delocalized Hartree-Fock limit: the probability of finding two electrons with the opposite spins on each atom is 1/4. The fifth term, quadratic in U, is the first deviation from the delocalized Hartree-Fock. It comes from a virtual excitation of two electrons in the bonding state, $(|a\rangle+|b\rangle)/\sqrt{2}$, to the antibonding state, $(|a\rangle-|b\rangle)/\sqrt{2}$, by the electron-electron interaction. The matrix element of such a two-electron excitation is $U/2$ and the excitation energy $(-4t)$, depressing the energy by $(U/2)^2(4t)^{-1}/2 = U^2/(32t)$. This term may be interpreted to express the effect of electron correlations reducing the probability of finding two electrons on the same atom.

The equilibrium distance r_2 between the two protons can be determined by minimizing the total energy, assuming that

$$t = t_0 \exp(-pr) , \tag{8.6}$$

$$R = R_0 \exp(-qr) , \tag{8.7}$$

where direct estimates show

$$2 < q/p < 3 . \tag{8.8}$$

We obtain the minimized energy and the equilibrium distance r_2 as follows

$$E_2/2 = E_0 + (1-p/q)t_2 + U/4 + (1+p/q)\, U^2/(32t) + \dots , \tag{8.9}$$

$$(q-p)r_2 = \ln[qR_0/(2p|t_0|)] + U^2/(32\, t_2{}^2) + \dots , \tag{8.10}$$

where

$$t_2 = t_0 \exp(-pr_2) . \tag{8.11}$$

In (8.9, 10) we see that the correlation term in U^2 stabilizes the dimer and increases the bond length.

In a similar way, the energy of the aggregate consisting of N atoms with one s electron surrounded by n equidistant nearest neighbors, in the expansion in powers of $|U/t|$, is obtained as follows [8.6]

$$E_N/N \simeq E_0 + nR/2 - w/4 + U/4 - U^2/(16w) + O(U^3/w^2), \tag{8.12}$$

where w is the effective s band-width approximately given by

$$w \simeq \sqrt{12\,n}\,|t|\,. \tag{8.13}$$

These expressions have numerical factors deduced from a rectangular band with a constant density of states. However, we can show that these factors are well adapted to more realistic band shapes. The band term $-w/4$ comes from the fact that the band shape is symmetrical in energy, and its lower half is occupied, yielding an average energy per electron of $-w/4$. The increase of w with $\sqrt{n}$ is equivalent to computing the kinetic energy assuming a Brownian motion of the electron from site to site. The term $U/4$ is the same as in the case of a dimer. The last correlation term $U^2/(16w)$ takes into account virtual excitation of two electrons from the occupied into the unoccupied part of the band with the average excitation energy w and the probability $U^2/16$.

Assuming again the variations of t and R, as given in (8.6, 7), respectively, one obtains the minimized energy and the equilibrium distance r_N as follows

$$E_N/N = E_0 - (1-p/q)w_N + U/4 - (1+p/q)U^2/(16w_N) + \dots\,, \tag{8.14}$$

$$(q-p)r_N = \ln[\sqrt{n/3}\,qR_0/(p|t_0|] + U^2/(4w_N{}^2) + \dots\,, \tag{8.15}$$

where

$$w_N = \sqrt{12\,n}\,|t_N|\,, \tag{8.16}$$

$$t_N = t_0\exp(-pr_N)\,. \tag{8.17}$$

Comparison of (8.10) with (8.15) shows that, for small values of $|U/t|$, the close-packed aggregate has the equilibrium distance somewhat larger than the dimer, but the correlation correction expands the dimer more than the close-packed aggregate. Further, comparision of (8.5) with (8.12), or that of (8.9) with (8.14), shows that qualitatively the Hartree-Fock term due to delocalization stabilizes the close-packed aggregate, while the correlation term favors the dimer. The repulsive term also favors the dimer.

These comparisons tell us that there is a critical value, $|V_c/t_2|$, below which a close-packed aggregate is more stable than a collection of dimers. The critical value depends upon the value of q/p in a critical way, as shown in Table 8.2. For the reasonable value of $q/p \simeq 2.5$, the critical value of $|U_c/t_2|$ is around 2, that makes valid the expansion in (8.5).

The conclusion seems in agreement with the fact that dimers are preferred by elements such as H, O, N and halogens having strong electronegativity while close-packed aggregates by all other elements such as alkali atoms. Note that dimers are usually weakly coupled by dispersion forces, which are out of our consideration because of the use of a single s orbital for each atom.

Table 8.2. Critical Values of $|U/t_2|$

q/p	2	2.5	3
$|U_c/t_2|$	0	2.1	3.5

8.3.2 Trimers of Monovalent Elements

The calculation of the total energy of a H_2 dimer by using the simplest Hubbard model, as given in the previous subsection, can be extended to a H_3 trimer [8.4]. The energies of (a) a dimer plus an atom, E_3^a, a linear chain, E_3^b, and an equilateral triangle, E_3^c, in the expansion in powers of $|U/t|$ are obtained as follows:

$$E_3^a/3 = E_0 + R/3 + 2t/3 + U/6 + U^2/(48t) + \dots, \tag{8.20}$$

$$E_3^b/3 = E_0 + 2R/3 + 2\sqrt{2}\, t/3 + 5U/24 + U^2/(51.9\, t) + \dots, \tag{8.21}$$

$$E_3^c/3 = E_0 + R + t + 2U/9 + U^2/(54\, t) + \dots \tag{8.22}$$

The geometrical configurations considered here are illustrated in Fig.8.5. In these energies, the band terms in t come from the one-electron energy levels (Fig.8.6).

The energy expressions in (8.20-22) show that the delocalization term in t tends to stabilize the equilateral triangle, but the terms in U and U^2/t tend to stabilize the dimer plus an atom. Therefore, there should be a critical value of $|U/t|$, below which the equilateral triangle is preferred and above which the dimer plus an atom is preferred. Since the repulsive term in R favors the dimer plus an atom. This critical value should increase with the ratio q/p. More detailed investigations reveal that for q/p = 2.5 the critical value of $|U/t|$ is expected to be small, the configuration of an equilateral triangle seems to be preferred. However, inspection of Fig.8.6 c reveals that the ground state of the triangle trimer is doubly degenerate in the absence of U, being subject to the Jahn-Teller distortion: the equilateral triangle may be distorted into an isocele triangle with the summit angle, $\theta \neq \pi/3$, as depicted in Fig.8.5 d. Treating $(\theta - \pi/3)$ as a small perturbation, one

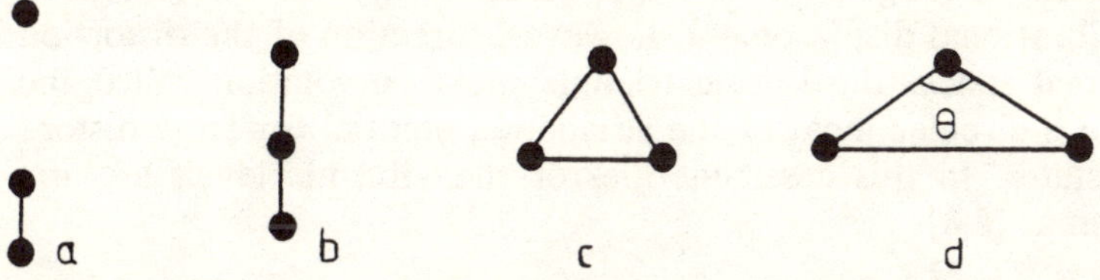

Fig.8.5. Various geometrical configurations of a trimer: (a) a dimer plus an atom or an ion; (b) a linear chain; (c) an equilateral triangle; (d) an isocele triangle [8.4]

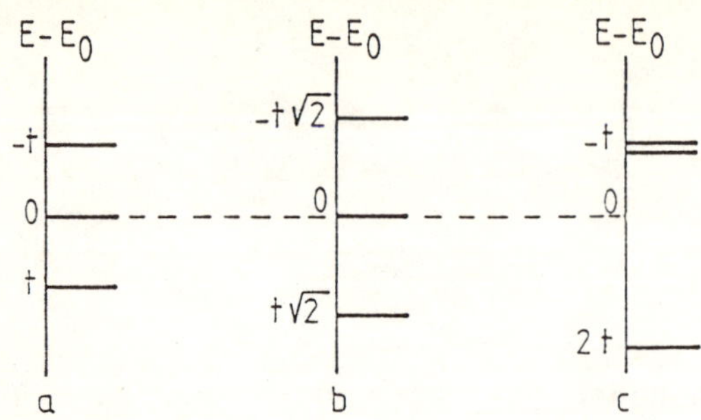

Fig.8.6. One-electron energy levels of a H_3 trimer in the absence of electron correlations: (**a**) a dimer plus an atom; (**b**) a linear chain; (**c**) an equilateral triangle [8.4]

can show that the equilibrium configuration is given by $\theta > \pi/3$. The critical angle depends upon the ratio q/p and the value of $|U/t|$. For q/p = 2.5 and vanishingly small value of U, one finds that the critical angle θ_c falls in the range between 60° and 90°.

8.3.3 Pseudorotation of Na_3

A beautiful experiment revealing the dynamical Jahn-Teller effect in the doubly-degenerate electronic excited state of a triangular Na_3 cluster was performed by means of resonant two-photon ionization spectroscopy [8.7]. Sodium clusters are produced by coexpansion of sodium vapor of moderate partial pressure together with argon gas through a nozzle. A high seeding ratio provides us with a low effective temperature for rotational and vibrational excitations in the ground-electronic state. Excitation by the first tunable laser brings the system from a lowest vibronic level of the ground state to vibronic levels of an electroncially excited state which is doubly-degenerate as in the ground state. The electron thus excited is then ejected by successive excitation by the second laser with a fixed frequency. The photoionized cluster, Na_3^+, is detected with a quadrupole mass spectrometer. Measurements are performed by setting the mass spectrometer on the mass peak of Na_3^+,while the ion signal is recorded as a function of the excitation wavelength of the first tunable laser. The observed spectrum is exhibited in Fig.8.7. The expanded spectrum of Fig.8.7 b can be considered to reflect vibronic levels of the doubly-degenerate excited electronic state of Na_3.

It is well known that a doubly-degenerate electronic state couples with a doubly-degenerate vibrational mode, giving the Jahn-Teller effect. In our case of a triangular Na_3, the atomic framework of an equilateral triangle originally assumed is distorted into an isocele triangle, as discussed in the previous subsection. As long as we take into account only the vibronic coupling linear in vibrational displacement, however, direction of the distortion is not fixed in real space: the isocele triangle makes a rotation called the pseudorotation in a circular moat of the adiabatic potential for large distortions and deep states. In this case, energies of the vibronic levels are approximately given as [8.8]

$$E_{nj} = (n+\tfrac{1}{2})\omega_0 + Aj^2 \quad (n = 0,1,2,...;\ j = \pm 1/2, \pm 3/2, \pm 5/2, ...) \qquad (8.23)$$

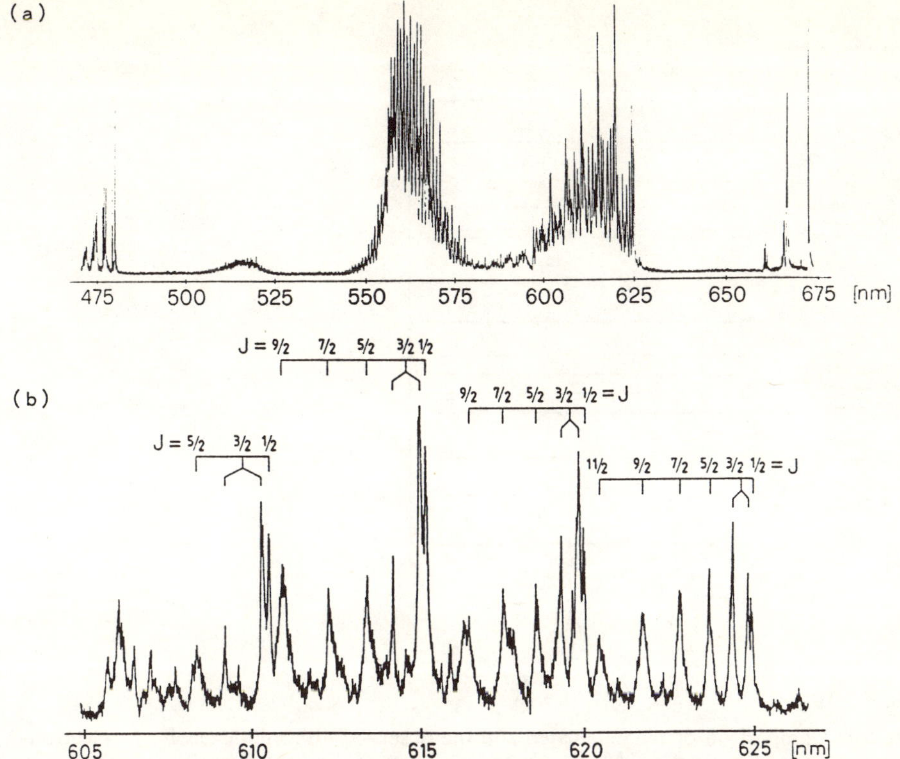

Fig.8.7. (**a**) Resonant two-photon ionization spectrum of Na_3. The wavelength is that of a tunable laser bringing Na_3 to electronically excited states. (**b**) Expanded spectrum of the region 600÷625 nm. Half-integral numbers are the quantum numbers labelling states of the internal pseudorotation in the excited state of Na_3 [8.7]

where ω_0 is the frequency of the doubly-degenerate vibration and $A = [2(\alpha/\omega_0^2)^2]^{-1}$, α being a constant of the linear vibronic coupling. The energy levels given by (8.23) are displayed in Fig.8.8 as LJT.

If we further take into account the vibronic-coupling cubic in the vibrational displacement, the moat bottom of the adiabatic potential becomes to show a sinusoidal oscillation with three minima and three maxima (saddle points). The calculated vibronic levels in this case are also exhibited in Fig.8.8 as CJT. The points indicated in the same figure represent the observed peaks. Agreement between the theory and the experiment looks very good.

8.4 Van der Waals and Metallic Mercury Clusters

It is of central interest for solid-state physicists to know how the occupied atomic orbitals containing valence electrons are broadened to form the band structure of a solid. From this view point, divalent metals are particularly

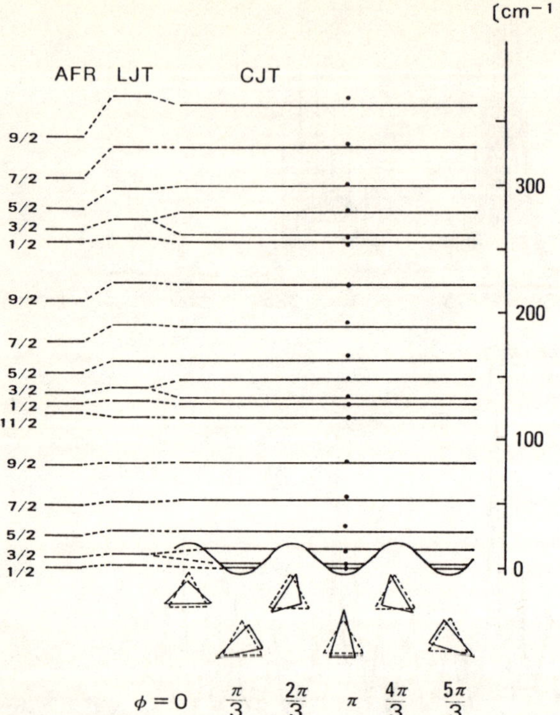

Fig.8.8. Comparison of the calculated and the observed vibronic levels in the electronically excited state of Na_3. The points represent the observed peaks. The AFR levels are those of free rotation in the absence of any vibronic coupling; LJT represent the calculated vibronic levels in the presence of a linear Jahn-Teller coupling; CJT are those in the presence of both the linear and cubic coupling. The sinusoidal curve at the bottom is the lowest adiabatic energy curve along the pseudorotation coordinate, and, below that, molecular geometries are given at energy minima and maxima (saddle points) [8.7]

interesting whose component atoms have an s^2 closed-shell configuration, as they can simply be considered to be insulators in the bulk. Contrary to such a simple consideration, metallic characters of these divalent metals come from the overlap between the filled s and the empty p bands. To probe the evolution from the van der Waals to the metallic bonding in divalent metal systems, we examine the change in the electronic structure of metal clusters as their size increases. As a divalent metal, we adopt mercury. The change in the electronic structure of mercury clusters is examined by measuring the photoionization due to core-valence transitions as a function of the photon energy [8.9].

Mercury clusters are formed in an adiabatic expansion of a pure mercury vapor through an appropriate nozzle. The light coming from the undulator of a storage ring, providing more than 10^{12} photons $(\text{Å}\cdot\text{s})^{-1}$, is focused at right angles to the neutral cluster beam. Then, the photoionized clusters are mass selected by a quadrupole mass spectrometer. The Photo-

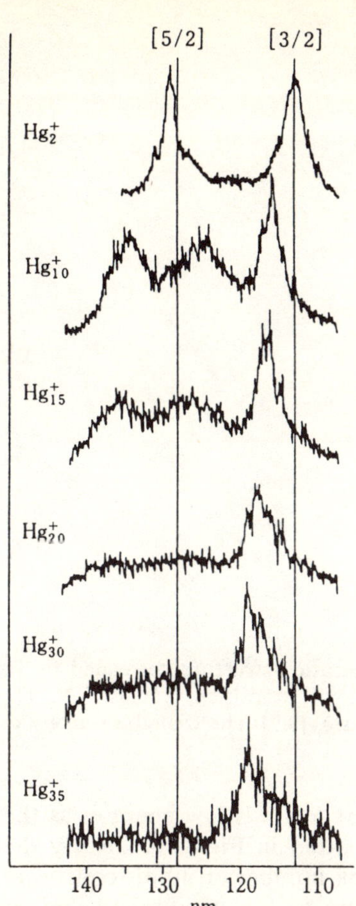

Fig.8.9. The recorded photoionization efficiency (PIE) curves for mercury clusters in the energy range of 8.5÷11.5 eV. The vertical straight lines represent atomic positions [8.9]

Ionization Efficiency (PIE) curve of each mass-selected cluster ion is displayed in Fig.8.9 for the cluster size of N = 2, 10, 15, 20, 30 and 35 in the energy range of 8.5÷11.5 eV.

In this energy range, we expect the atomic excitation,

$$\mathrm{Hg}(5d^{10}\,6s^2\ {}^1S_0) \rightarrow \mathrm{Hg}^*[5d^9\,6s^2\,(D_{5/2},D_{3/2})6p]\ , \tag{8.24}$$

and the succeeding ejection of an Auger electron (autionization),

$$\mathrm{Hg}^*(5d^9\,6s^2\,6p) \rightarrow \mathrm{Hg}^+(5d^{10}\,6s) + e^-\ . \tag{8.25}$$

For the atom, only the $D_{3/2}$ state leads to the autoionization, whereas both the $D_{3/2}$ and $D_{5/2}$ states induce the autoionization for Hg_N with $N \geq 2$. The positions of two peaks of the PIE curve corresponding to the autoionization induced by the $D_{3/2}$ and $D_{5/2}$ excited states of the atom are indicated by the vertical straight lines denoted as [3/2] and [5/2] in Fig.8.9.

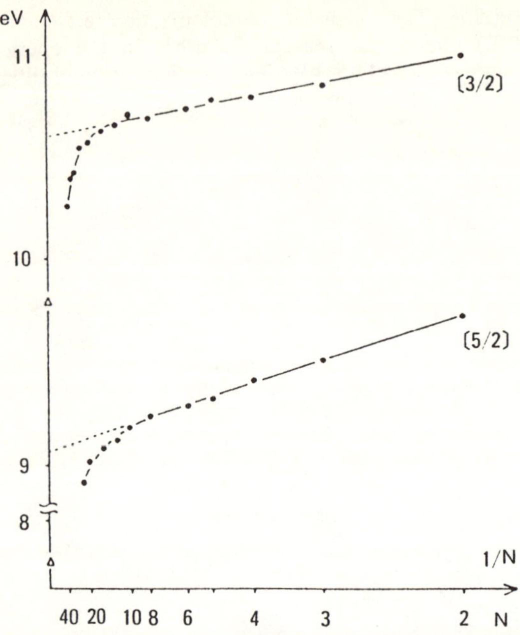

Fig.8.10. The observed peak energies of the autoionization structure correlated to the $Hg(5d^{10}\ 6s^2\ {}^1S_0) \rightarrow Hg^*[(5d^9\ 6s^2)\ D_{5/2,3/2}\ 6p]$ atomic transitions versus 1/N. The straight lines illustrate the 1/N behavior for small masses [8.9]. The triangles for $N = \infty$ are taken from [8.10]

In Fig.8.9 one sees that the two peaks shift towards low energies as the cluster size increases. The observed shift is plotted in Fig.8.10. Clearly the shift for $N \leq 12$ is inversely proportional to the number of atoms constituting the cluster. This dependence is analogous to the excitonic behavior which is characteristic of van der Waals systems [8.9]. For $N > 12$ the 1/N dependence of the shift is no longer observed indicating deviation from the van der Waals bonding in larger clusters.

For $N > 20$ only the 3/2 peak is observed (Fig.8.9). The peak is broadened and its spectral line shape becomes asymmetric. Since the corresponding absorption in the bulk exciting a 5d core electron to the unfilled part of the sp band shows a step-like spectral shape extending towards high energies, the observed 3/2 autoionization line with a remarkable asymmetry seems to indicate the onset of a gradual construction of the sp band at $N \sim 20$. It is suggested that the transition from van der Waals ($N \leq 12$) to metallic mercury clusters takes place over a wide range of the cluster size [8.9].

8.5 Prospects of Microcluster Research

In the first chapter, it was mentioned that one origin of microcluster research may be found in the study of *fine particles* consisting of 10^3 - 10^5 atoms. However, the physics relevant for research on fine particles is quite different from that for microscluster research. The difference comes from the fact that microclusters of a given shape and a size can be extracted and their properties measured, but this kind of extraction is impossible for fine particles. Only a statistical treatment is possible for fine particles of fixed size but different shapes forming a large ensemble: these fine particles are nearly degenerate in energy.

Contrary to the above statement, the recent development of a cluster source and mass spectrometry has made it possible to observe some oscillatory features in the abundance of Na_N clusters with N up to $\simeq$20000 as shown in Figs.8.11 and 12 [8.11]. In Fig.8.11, two sequences of the oscillatory structures, minima of the counts, are observed: one ends at N = 1430 and another starts at N = 1980. The first sequence is interpreted to be due to closing the shell of valence electrons characterized by approximate quantum number $3n+\ell$, where n denotes the number of nodes in the solution to the radial Schrödinger equation and ℓ is the angular momentum for the system with a spherical potential. The second sequence is interpreted to be due to closing the atomic shell of a cuboctahedron or an icosahedron, which contains

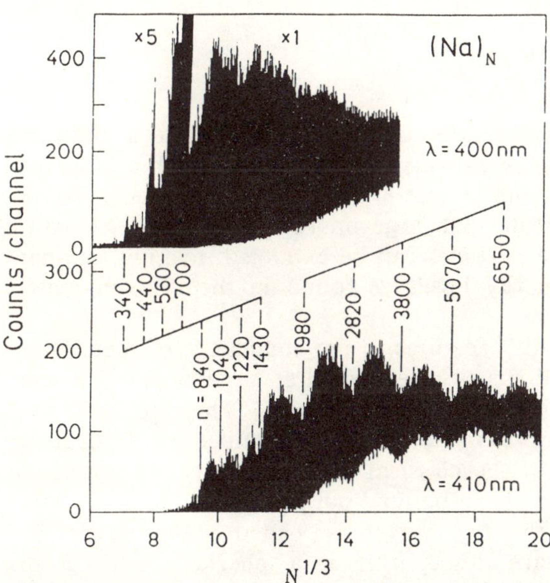

Fig.8.11. Mass spectras of Na_N clusters photoionized with 400 and 410 nm light. The vertical axis denotes the total number of counts accumulated in a 40 ns time channel after about 10^5 laser shots. Two sequences of structures are observed at equally spaced intervals on the $N^{1/3}$ scale [8.11]

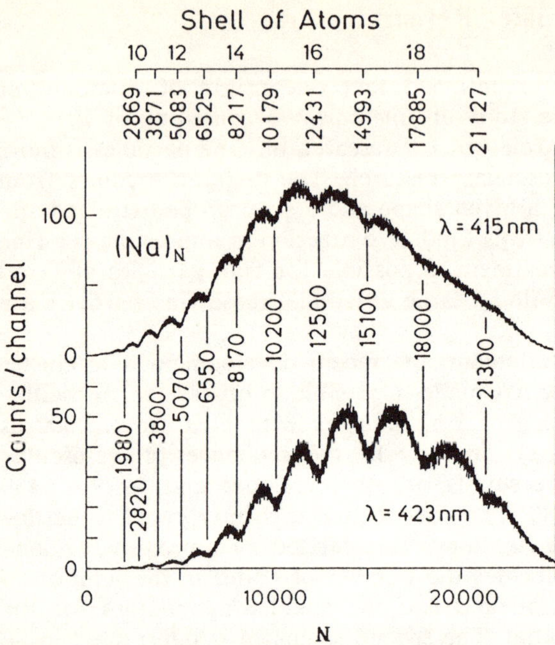

Fig.8.12. Averaged mass spectra of Na_N clusters photoionized with 415 and 423 nm light. Well-defined minima occur at values of N corresponding to the total number of atoms in icosahedra with closed atomic shells, which are indicated at the top [8.11]

$$I_K = \frac{1}{3}(10K^3 - 15K^2 + 11K-3)$$

atoms when its Kth shell is closed. Although there still remains a difficulty in answering the question how sensitively the stability of such a large and metallic cluster depends upon the atomic structure of the surface, the results in Figs.8.11 and 12 indicate that large microclusters of an icosaedral shape in the size range of fine particles can be extracted, making redunant the statistical treatment of energy levels as found in the random-matrix theory.

Modulation of the oscillatory features in the abundance of metal clusters due to the electronic shell structure expected at large N has also been discussed [8.12]. The modulation is called a *supershell.* The electronic shell structure characterized by approximate quantum number $3n+\ell$ for large clusters may be identified with the classical motion in a closed triangular orbit in a spherical cavity. It has been pointed out [8.13] that the modulation comes from interference of the two waves associated with semiclassical motions in triangular and square closed orbits in a spherical cavity. Actually, the shell correction U in (3.22) for a spherical potential of the Wood-Saxon type is calculated as shown in Fig.8.13 [8.12], where minima in the oscillation amplitude at N = 1000 and 4000 and maxima at N = 500 and 2500 are due to the supershell structure in the level density.

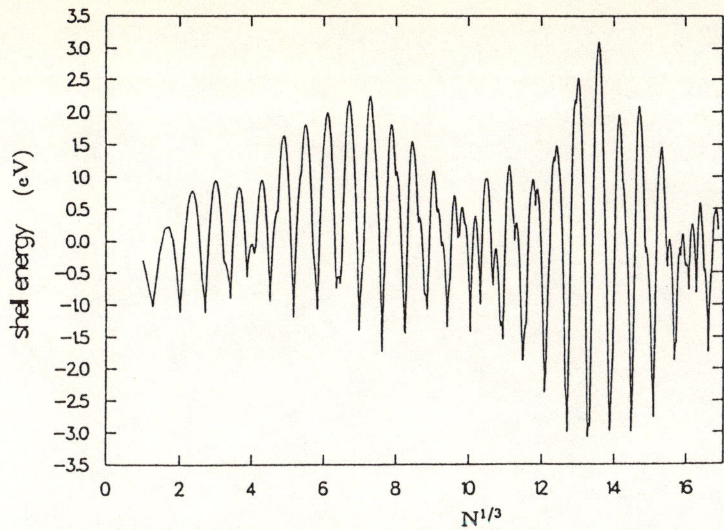

Fig.8.13. The shell correction, the shell oscillating part of the total energy of valence electrons, calculated for the Woods-Saxon potential as a function of $N^{1/3}$. The minima of the envelope of the shell correction at N ≃ 1000 and 4000 are due to the supershell structure [8.12]

The fact that the predicted supershell has not been confirmed experimentally yet has left us with an interesting problem to be solved in the future. The onset of the second sequence of the observed oscillatory structure (Fig.8.11) at size N, where the first node of the supershell is predicted, may be accidental. This recent experimental and theoretical microcluster research indicates that fundamental knowlege about the transition from microclusters to the corresponding bulk is still lacking.

One difficulty in microscluster research is the problem of accumulating enough microclusters of a given size for the measurement of their properties. A conceivable way of overcoming such a difficulty is to develop highly sensitive mesasurement techniques. Spectrosopic studies along this line are developing rapidly [8.13, 14] and will constitute a central area of physical studies of microsclusters in the near future. Another way of overcoming the difficulty is to accumulate microclusters in or on solid matrices. However, even by this method, it is not easy to produce clusters of a selected size. Moreover, there is a fundamental problem arising from cluster-matrix interactions. These interactions are difficult to determine without prior knowledge of the noninteracting free clusters. For this reason, experimental studies of concentrated cluster beams using detection methods of high sensitivity should be further developed.

References

Chapter 1

1.1 R. Kubo: J. Phys. Jpn. **17**, 975 (1962)

1.2 L.P. Gor'kov, G.M. Eliashberg: Zh. Eksp. Theor. Fiz. **48**, 1407 (1965) [Sov. Phys. JETP **21**, 940 (1965)]

1.3 C.E. Porter (ed.): *Statistical Theories of Spectra: Fluctuations* (Academic, New York 1965)

1.4 R. Denton, B. Möhlschlegel, D.J. Scalapino: Phys. Rev. Lett. **26**, 707 (1971); Phys. Rev. B **7**, 3589 (1986)

1.5 S. Tanaka, S. Sugano: Phys. Rev. B **34**, 740 (1986)

1.6 W.D. Knight, K. Clemenger, W.A. de Heer, W.A. Saunders, M.Y. Chou, M.L. Cohen: Phys. Rev. Lett. **52**, 2141 (1984)

1.7 I. Katakuse, T. Ichihara, Y. Fujita, T. Matsuo, T. Sakurai, H. Matsuda: Int. J. Mass. Spectrom. and Ion Proc. **67**, 229 (1985)

1.8 W.A. de Heer, W.D. Knight, M.Y. Chou, M.L. Cohen: Electronic shell structure and metal clusters. *Solid State Physics* **40**, 93–181 (Academic, New York 1987)

1.9 M.G. Mayer, J.H. Jensen: *Elementary Theory of Nuclear Shell Structure* (Wiley, New York 1955)

1.10 M.R. Hoare: Adv. Chem. Phys. **40**, 49 (1979)

Chapter 2

2.1 J. Farges, M.F. de Feraudy, B. Raoult, G. Torchet: J. Chem. Phys. **78**, 5067 (1983); ibid. **84**, 3491 (1986)

2.2 S. Iijima: In *Microclusters*, ed. by S. Sugano, Y. Nishina, S. Ohnishi. Springer Ser. Mat. Sci., Vol.4 (Springer, Berlin, Heidelberg 1987) p.186

2.3 M.R. Hoare: Adv. Chem. Phys. **40**, 49 (1979)

2.4 M. Mitome, Y. Tanishiro, K. Takayanagi: Z. Phys. D **12**, 45 (1989)

2.5 J. Jellinek, T.L. Beck, R.S. Berry: J. Chem. Phys. **84**, 2783 (1986)
D.W. Heermann: *Computer Simulation Methods in Theoretical Physics*, 2nd edn. (Springer, Berlin, Heidelberg 1990)

2.6 R.S. Berry, J. Jellinek, G. Natanson: Phys. Rev. A **30**, 919 (1984)

2.7 S. Sawada, S. Sugano: Z. Phys. D **14**, 247 (1989)

2.8 R.P. Gupta: Phys. Rev. B **23**, 6265 (1981)

2.9 M.B. Gordon, F. Cyrot-Lackmann, M.C. Desjonqueres: Surf. Sci. **80**, 159 (1979)

2.10 D. Tomanek, S. Mukherjee, K.H. Bennemann: Phys. Rev. B **28**, 665 (1983)

2.11 E.M. Pearson, H. Halicioglu, W.A. Tiller: Phys. Rev. A **32**, 3030 (1985)

2.12 K. Binder, D. Stauffer: *Application of the Monte Carlo Method in Statistical Physics* ed. by K. Binder, 2nd edn., Topics Current Phys. Vol.36 (Springer, Berlin, Heidelberg 1987) p.1

2.13 K. Binder, D.W. Heermann: *Monte Carlo Simulation in Statistical Physics*, Springer Ser. Solid-State Sci., Vol.80 (Springer, Berlin, Heidelberg 1988)

2.14 H.T. Diep, S. Sawada, S. Sugano: Phys. Rev. B **39**, 9252 (1989)
2.15 N. Quirke: Mol. Simulation **1**, 249 (1988)
2.16 J.P. Hansen, L. Varlet: Phys. Rev. **184**, 151 (1969)
2.17 T.L. Beck, J. Jellinek, R.S. Berry: J. Chem. Phys. **87**, 545 (1987)
2.18 T.L. Beck, R.S. Berry: J. Chem. Phys. **88**, 3910 (1988)
2.19 N. Fujima, T. Yamaguchi: J. Phys. Soc. Jpn. **58**, 3290 (1989)

Chapter 3

3.1 W.D. Knight, K. Clemenger, W.A. de Heer, W.A. Saunders, M.Y. Chou, M.L. Cohen: Phys. Rev. Lett. **52**, 2141 (1984)
3.2 I. Katakuse: In *Microclusters*, ed. by S. Sugano, Y. Nishina, S. Ohnishi, Springer Ser. Mat. Sci., Vol.4 (Springer, Berlin, Heidelberg 1987) p.10
3.3 M.G. Mayer, J.H. Jensen: *Elementary Theory of Nuclear Shell Structure* (Wiley, New York 1955)
3.4 J.R. Smith (ed.): *Theory of Chemisorption*, Topics Curr. Phys., Vol.19 (Springer, Berlin, Heidelberg 1980), Chap.1
3.5 M.Y. Chou, A.N. Cleland, M.L. Cohen: Solid State Comm. **52**, 645 (1984)
3.6 B.K. Rao, P. Jena, M. Manninen: Phys. Rev. B **32**, 477 (1985)
B.K. Rao, P. Jena: Phys. Rev. B **32**, 2058 (1985)
3.7 Y. Ishii, S. Ohnishi, S. Sugano: Phys. Rev. B **33**, 5271 (1986)
3.8 A. Tamura, T. Ichinokawa: Surf. Sci. **136**, 437 (1984)
3.9 V.N. Bogomolov, N.A. Klushin, N.M. Okuneva, E.L. Plachenova, V.I. Pogrebnoi, F.A. Chudnouskii: Soviet Phys. - Solid State **13**, 1256 (1971)
3.10 V.R. Pandharipande, S.C. Pieper, R.B. Wiringa: Phys. Rev. B **34**, 4571 (1986)
3.11 P. Hohenberg, W. Kohn: Phys. Rev. **136**, B 864 (1964)
3.12 W. Kohn, L.J. Sham: Phys. Rev. **140**, B 1133 (1965)
3.13 J.P. Perdew, A. Zunger: Phys. Rev. B **23**, 5048 (1981)
3.14 W.A. de Heer, W.D. Knight, M.Y. Chou, M.L. Cohen: Electronic shell structure and metal clusters. *Solid State Physics* **40**, 93-181 (Academic, New York 1987)
3.15 A. Herrmann, E. Schumacher, L. Wöste: J. Chem. Phys. **68**, 2327 (1978)
3.16 D.M. Wood: Phys. Rev. Lett. **46**, 749 (1981)
3.17 N.W. Ashcroft, D.C. Langreth: Phys. Rev. **155**, 682 (1967)
3.18 N.W. Ashcroft: J. Phys. C **1**, 232 (1968)
3.19 Y. Ishii, S. Sugano: In *Microclusters* ed. by S. Sugano, Y. Nishina, S. Ohnishi, Springer Ser. Mat. Sci., Vol.4 (Springer, Berlin, Heidelberg 1987) p.17
3.20 A. Bohr, B.R. Mottelson: Mat. Fys. Medd. Dan. Vid. Selsk. **27**, No.16 (1953)
3.21 V.M. Strutinsky: Nuclear Phys. A **95**, 420 (1967); ibid. A **122**, 1 (1968)
3.22 L.D. Landau, E.M. Lifshitz: *Electrodynamics of Continuous Media* (Pergamon, London, 1960) p.20
3.23 S.G. Nilsson: Mat. Fys. Medd. Dan. Vid. Selsk. **29**, No.16 (1955)
3.24 K. Clemenger: Phys. Rev. B **32**, 1359 (1985)
3.25 M. Brack, J. Damgaard, A.S. Jensen, H.C. Pauli, V.M. Strutinsky, C.Y. Wong: Rev. Mod. Phys. **44**, 320 (1972)
3.26 M. Nakamura, Y. Ishii, A. Tamura, S. Sugano: Phys. Rev. A **42**, 2267 (1990)
S. Sugano, A. Tamura, Y. Ishii: Z. Phys. D **12**, 213 (1989)
3.27 R. Balian, C. Bloch: Ann. Phys. **60**, 401 (1970)
3.28 H. Weyl: Nachr. Akad. Wiss. Göttingen (1911) p.110

Chapter 4

4.1 J.L. Martins, J. Buttet, R. Car: Phys. Rev. B **31**, 1804 (1985)
4.2 C. Satoko: Phys. Rev. B **30**, 1754 (1984)
4.3 A. Herrmann, E. Schumacher, L. Wöste: J. Chem. Phys. **68**, 2327 (1978)
4.4 K.I. Peterson, P.D. Das, R.W. Farley, A.W. Castleman, Jr.: J. Chem. Phys. **80**, 1780 (1984)
4.5 J.L. Martins, R. Car, J. Buttet: Surf. Sci. **106**, 265 (1981)
4.6 D.M. Wood: Phys. Rev. Lett. **46**, 749 (1981)
4.7 E. Schumacher, M. Kappes, K. Marti, P. Radi, M. Shar, B. Schmidhalter: Ber. Bunsenges. Phys. Chem. **88**, 220 (1984)
4.8 G.A. Thompson, F. Tischler, D.M. Lindsay: J. Chem. Phys. **78**, 5946 (1983)
4.9 D.M. Lindsay, G.A. Thompson: J. Chem. Phys. **77**, 1114 (1982)
4.10 G.A. Thompson, D.M. Lindsay: J. Chem. Phys. **74**, 959 (1981)
4.11 N. Fujima, T. Yamaguchi: J. Phys. Soc. Jpn. **58**, 1334 (1989)
4.12 H. Tatewaki, E. Miyoshi, T. Nakamura: J. Chem. Phys. **76**, 5073 (1982)
4.13 B. Delley, D.E. Ellis, A.J. Freeman, E.J. Baerends, D. Post: Phys. Rev. B **27**, 2132 (1983).
4.14 C.L. Pettiette, S.H. Yang, M.J. Craycraft, J. Conceicao, R.T. Laaksonen, O. Cheshnovsky, R.E. Smalley: J. Chem. Phys. **88**, 5377 (1988)
4.15 N. Fujima, T. Yamaguchi: J. Phys. Soc. Jpn. **58**, 3290 (1989)
4.16 K. Lee, J. Callaway, K. Kwong, R. Tang, A. Ziegler: Phys. Rev. B **31**, 1796 (1985)
4.17 R.P. Messmer, S.K. Knudson, K.H. Johnson, J.B. Diamond, C.Y. Yang: Phys. Rev. B **13**, 1396 (1976)
4.18 M. Tomonari, H. Tatewaki, T. Nakamura: J. Chem. Phys. **85**, 2875 (1986)
4.19 D.M. Cox, D.J. Trevor, R.L. Whetten, E.A. Rohlfing, A. Kaldor: Phys. Rev. B 32, 7290 (1985)
4.20 I. Katakuse: In *Microclusters*, ed. by S. Sugano, Y. Nishina, S. Ohnishi, Springer Ser. Mat. Sci., Vol.4 (Springer, Berlin, Heidelberg 1987) p.10
4.21 M.F. Jarrold, J.E. Bower, J.S. Kraus: J. Chem. Phys. **86**, 3876 (1987)
4.22 D.M. Cox, D.J. Trevor, R.L. Whetten, E.A. Rohlfing, A. Kaldor: J. Chem. Phys. **84**, 4651 (1986)
4.23 T.H. Upton: J. Chem. Phys. **86**, 7054 (1987)
4.24 L. Hanley, S.A. Ruatta, S.L. Anderson: J. Chem. Phys. **87**, 260 (1987)

Chapter 5

5.1 E.A. Rohlfing, D.M. Cox, A. Kaldor: J. Chem. Phys. **81**, 3322 (1984)
5.2 H.W. Kroto, J.R. Heath, S.C. O'Brion, R.F. Curl, R.E. Smalley: Nature **318**, 162 (1985)
5.3 K. Raghavachari, J.S. Binkley: J. Chem. Phys. **87**, 2191 (1987)
5.4 L.A. Bloomfield, R.R. Freeman, W.L. Brown: Phys. Rev. Lett. **54**, 2246 (1985)
5.5 Q.-L. Zhang, Y. Liu, R.F. Curl, F.K. Tittel, R.E. Smalley: J. Chem. Phys. **88**, 1670 (1988)
5.6 J. Van de Walle, P. Joyes: Phys. Rev. B **32**, 8381 (1985)
5.7 G. Durand, J.P. Daudey, J.P. Malrieu: J. Physique **47**, 1335 (1986)
5.8 S. Saito, S. Ohnishi, C. Satoko, S. Sugano: J. Phys. Soc. Jpn. **55**, 1791 (1986)
5.9 S. Saito, S. Ohnishi, S. Sugano: Phys. Rev. B **33**, 7036 (1986)
5.10 K. Raghavachari: J. Chem. Phys. **84**, 5672 (1986)
5.11 S. Saito, S. Ohnishi, C Satoko, S. Sugano: Z. Physik D **14**, 237 (1989)
5.12 W. Kratschmer, L.D. Lamb, K. Fostiropoulos, D.R. Huffman: Nature **347**, 354 (1990)

Chapter 6

6.1 O. Echt, K. Sattler, E. Recknagel: Phys. Rev. Lett. **47**, 1121 (1981)
6.2 A.L. Mackay: Acta Crystallogr. **15**, 916 (1962)
6.3 I.A. Harris, R.S. Kidwell, J.A. Northby: Phys. Rev. Lett. **53**, 2390 (1984)
6.4 I.A. Harris, K.A. Norman, R.V. Mulkern, J.A. Northby: Chem. Phys. Lett. **130**, 316 (1986)
6.5 P.W. Stephens, J.G. King: Phys. Rev. Lett. **51**, 1538 (1983)
6.6 N. Kobayashi, T. Kojima, Y. Kaneko: J. Phys. Soc. Jpn. **57**, 1528 (1988)
6.7 V.R. Pandharipande, Z.G. Zabolitzky, S.C. Pieper, R.B. Wiringa, U. Helmbrecht: Phys. Rev. Lett. **50**, 1676 (1983)
6.8 M.H. Kalos, M.A. Lee, P.A. Whitlock, G.V. Chester: Phys. Rev. B **24**, 115 (1981)
6.9 V.R. Pandharipande, S.C. Pieper, R.B. Wiringa: Phys. Rev. B **34**, 4571 (1986)
6.10 R.P. Feynman, M. Cohen: Phys. Rev. **102**, 1189 (1956)

Chapter 7

7.1 H. Shinohara, N. Nishi: J. Chem. Phys. **83**, 1939 (1985)
7.2 U. Nagashima, H. Shinohara, N. Nishi, H. Tanaka: J. Chem. Phys. **84**, 209 (1986)
7.3 S.S. Lin: Rev. Sci. Instrum. **44**, 516 (1973)
7.4 J.L. Kassner, jr., D.E. Hagen: J. Chem. Phys. **64**, 1860 (1976)
7.5 O. Matsuoka, E. Clementi, M. Yoshimine: J. Chem. Phys. **64**, 1351 (1976)
7.6 I.P. Buffey, W.B. Brown: Chem. Phys. Lett. **109**, 59 (1984)
7.7 N.R. Kestner, J. Jortner: J. Phys. Chem. **88**, 3818 (1984)
7.8 C.E. Klots, R.N. Compton: J. Chem. Phys. **69**, 1644 (1978)
7.9 A.M. Sapse, L. Osorio, G. Snyder: Int. J. Quantum Chem. **26**, 223 (1984)
7.10 H. Haberland, C. Ludewigt, H.-G. Schindler, D.R. Worsnop: J. Chem. Phys. **81**, 3742 (1984)
7.11 M. Knapp, O. Echt, D. Kreisle, E. Recknagel: J. Chem. Phys. **85**, 636 (1986)
7.12 H. Haberland, C. Ludewigt, H.-G. Schindler, D.R. Worsnop: Phys. Rev. A **36**, 967 (1987)
7.13 M. Knapp, O. Echt, D. Kreisle, E. Recknagel: J. Phys. Chem. **91**, 2601 (1987)
7.14 B.J. Berne, D. Thirumalai: Ann. Rev. Phys. Chem. **37**, 401 (1986)
7.15 R.N. Barnett, U. Landman, C.L. Cleveland, N.R. Kestner, J. Jortner: J. Chem. Phys. **88**, 6670 (1988)
7.16 R.N. Barnett, U. Landman, C.L. Cleveland, J. Jortner: Phys. Rev. Lett. **59**, 811 (1987)
7.17 J.R. Reimers, R.O. Watts, M.L. Klein: Chem. Phys. **64**, 95 (1982)
J.R. Reimers, R.D. Watts: Chem. Phys. **85**, 83 (1984)
7.18 U. Landman, R.N. Barnett, C.L. Cleveland, D. Sharf, J. Jortner: J. Phys. Chem. **91**, 4890 (1987)
R.N. Barnett, U. Landman, C.L. Cleveland, J. Jortner: J. Chem. Phys. **88**, 4429 (1988)
7.19 M. Matsuzawa: In *Rydberg States of Atoms and Molecules*, ed. by R.F. Stebbings, F.B. Dunning (Cambridge Univ. Press, London 1983) p.267
7.20 T. Kondow, K. Mitsuke: J. Chem. Phys. **83**, 2612 (1985)
7.21 M. Tsukada, N. Shima, S. Tsuneyuki, H. Kageshima: J. Chem. Phys. **87**, 3927 (1987)
7.22 H. Sumi: J. Phys. Soc. Jpn. **49**, 1701 (1980)
7.23 H. Adachi, M. Tsukada, C. Satoko: J. Phys. Soc. Jpn. **44**, 1045 (1978)
7.24 D.G. Hopper: Chem. Phys. **53**, 85 (1980)

Chapter 8

8.1 P. Fayet, M.J. McGlinchey, L.H. Wöste: J. Am. Chem. Soc. **109**, 1733 (1987)
8.2 J.W. Lauher: J. Am. Chem. Soc. **100**, 5305 (1978)
8.3 P. Fayet, F. Granzer, G. Hagenbart, E. Moisar, B. Pischel, L. Wöste: Phys. Rev. Lett. **55**, 3002 (1985)
8.4 J. Friedel: *Microclusters*, ed. by S. Sugano, Y. Nishina, S. Ohnishi, Springer Ser. Mat. Sci., Vol.4 (Springer, Berlin, Heidelberg 1987) p.26
8.5 J.J. Hubbard: Proc. Roy. Soc. A**276**, 238 (1963)
8.6 J. Friedel: *Physics and Chemistry of Electrons and Ions in Condensed Matter*, ed. by J. V. Acrivos, N.F. Mott, A.D. Joffe, NATO ASI Ser. C **130** (Reidel, Dordrecht 1984)
8.7 G. Delacrétaz, E.R. Grant, R.L. Whetten, L. Wöste, J.W. Zwanziger: Phys. Rev. Lett. **56**, 2598 (1986)
8.8 H.C. Longuett-Higgins, U. Öpik, M.H.L. Pryce, R.A. Sack: Proc. Roy. Soc. (London) A**244**, 1 (1958)
8.9 C. Bréchignac, M. Broyer, Ph. Cahuzac, G. Delacretaz, P. Labastie, J.P. Wolf, L. Wöste: Phys. Rev. Lett. **60**, 275 (1988)
8.10 K. Bartschat, P. Scott: J. Phys. B**18**, L 191 (1985)
8.11 T.P. Martin, T. Bergmann, H. Gohlich, T. Lange: Chem. Phys. Lett. **172**, 209 (1990)
8.12 H. Nishioka, K. Hansen, B.R. Mottelson: Phys. Rev. B **42**, 9377 (1990)
8.13 R. Balian, C. Bloch: Ann. Phys. (NY) **69**, 76 (1971)
8.14 E. Recknagel, O. Echt (eds.): Proc. 5th Int'l Symp. on Small Particles and Inorganic Clusters (Springer, Berlin, Heidelberg 1991)
8.15 M.C. Cohen, W.D. Knight: Phys. Today p.42 (December 1990)

Subject Index

Printing: Mercedesdruck, Berlin
Binding: Buchbinderei Lüderitz & Bauer, Berlin